Just in Time Technology

Doing Better with Fewer

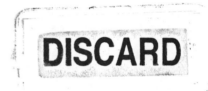

by Jamie McKenzie

FNO Press
Bellingham, Washington
http://fnopress.com
http://store.yahoo.com/fnopress/

FNO Press   http://fnopress.com

This work contains some articles previously published in a number of journals and publications.

"Smart Technology 1" first appeared as "Prospecting for Digital Resources" in the January/February, 2002 issue of **Multimedia Schools** and as an article in the March, 2002 issue of **FNO**.

"Smart Technology 2" first appeared as "Tech Smart: Making Discerning Technology Choices" in the March 2002 issue of **Multimedia Schools** and as an article in the May, 2002 issue of **FNO**.

"Post Deployment Action Plan"- **FNO,** November/December, 2002

"Why Not Share?" - **FNO,** April, 2002.

"Creating Good New Ideas 1"- **FNO,** June, 2001.

"Creating Good New Ideas 2" - **FNO,** September, 2002 .

"Paper Still Works" - **FNO,** November/December, 2001.

"The Medium is Not the Literacy" - **FNO,** June, 2002.

"Beware the Visionary" - **FNO,** April, 2002.

"After Laptop" - **FNO,** April, 2002.

"Stories of Adult Learning" first appeared as "Beyond Toolishness" in the September, 2002 issue of **Multimedia Schools**.

"Verity - the Search for Difficult Truth" - **FNO,** May, 2001.

"Going Unplugged" - **FNO**, October, 2001.

Just in Time Technology is dedicated to those who possess the courage to point out when the emperor has no clothes or the technology program has no clear sense of purpose.

This book is also dedicated to my wife and best friend, Gretchen, whose loving approach to teaching young ones (with or without technologies) is an inspiration.

About the author . . .

Jamie McKenzie is the Editor of **From Now On - The Educational Technology Journal**, a Web-based zine published since 1991. In this journal he has argued for information literate schools. More than twenty-five per cent of his 22,000+ subscribers live in countries outside of North America such as Australia, New Zealand, Sweden, Malaysia, and Singapore.

From 1993-1997, Jamie was the Director of Libraries, Media and Technology for the Bellingham (WA) Public Schools, a district of 19 schools and 10,000 students that was fully networked with 2000 desktops all tied to the Internet in 1995. He has since moved on to support literacy, technology planning, and professional development across North America, Australia, and New Zealand.

A graduate of Yale with an M.A. from Columbia and Ed.D. from Rutgers, Jamie has been a middle school teacher of English and social studies, an assistant principal, an elementary principal, an assistant superintendent in Princeton (NJ), and a superintendent of two districts on the East Coast of the States. He also taught four-year-olds in Sunday school.

Jamie has published and spoken extensively on the introduction of new technologies to schools. In recent times, he has paid particular attention to what he calls discerning uses of new technologies.

A full resume listing publication credits and a detailed career history is available online at http://fno.org/JM/resume.html.

Introduction

We may not need a laptop sitting on every student's desktop poised for that moment when a computer just might be the right tool for the job.

This book was designed for teachers, students, parents and principals in schools where sharing, planning and restraint are still cherished social values - where resources are limited and thrift is yet considered a virtue.

This book is dedicated to the notion that thoughtful planning can lead to important cost savings along with optimal use of new tools - thereby freeing funds for other goals, activities and technologies.

Emerging from a decade of exuberance, foolishness, greed, exaggerated vendor claims, fraudulent accounting and a frenzied overemphasis upon digital tools and life-styles, schools are learning to be discerning consumers, approaching new tools and gadgets with reasoned skepticism and judicious thinking.

Most of us reject the notion that this is a digital world - preferring, instead, a much richer mix of inputs and resources. We know the joys of going unplugged, walking along a beach, and curling up with a musty old book like **Treasure Island** - a family heirloom passed down through several generations complete with original illustrations by N. C. Wyeth.

Just in Time suggests ways to move equipment around to optimize results and maximize a school's returns on investments.

The whole notion of "anytime, anywhere" deployment now seems outmoded, wasteful and thoughtless. It is a marketing slogan of a bubblegum economy that surged on speculation but burst on bottom lines. We now see the wisdom of putting profligate ways behind us.

A French king promised a chicken in every pot. Some of today's politicians suggest a laptop for every child - not to mention cell phones and personal handheld digital assistants.

The road to perdition is paved, it seems, with the skeletons of broken technology promises - in the world of high finance and business as well as the world of schools.

We cannot afford to invest in companies without business plans or in technologies without credible learning plans and prospects. Schools are under considerable pressure to create waves of thinking, problem-solving students capable of mastering the demanding skills expected by provincial and state governments across North American, Australia, and elsewhere.

The essential learnings are not the spreadsheeting, powerpointing and mouse movements of a technology-driven program or culture. They are, instead, the communication, problem-solving, decision-making and teaming that good schools have stressed for half a century or more. This is no brave new world, corporate agenda. These are classical goals - to raise young ones capable of thinking independently and working productively as part of a caring community with whatever tools make sense for the jobs at hand.

Bamboozlement is not new in this century, but recent corporate failures, excesses, bankruptcies and frauds have raised our awareness of deceptive business practices to a high level. Schools have tasted some of these misleading business practices and learned to challenge and question more aggressively.

Demagogues employed bamboozlement in the previous century, offering simple solutions to complex issues. Snake oil sales folk have been around for a long time. But the addition of some accounting firms to the list of suspects is a sad development as we have long counted on these companies to vouchsafe the integrity of corporate and government financial reports.

Schools must view the technology planning process within this broader context of economic boom, bust and speculation, recognizing that the rush to network classrooms was based on little data and much hope and dazzle. Many people made lots of money selling tools that have yet proven worthy.

The best uses of new technologies often entail a thoughtful blending of those tools with more classical tools. We must view the exaggerated claims and promises of vendors and entrepreneurs with informed skepticism and discernment. The creation of program should be driven by learning goals shaped by the sagacity and experience of educators who maintain proximity to classrooms and children.

This is a time to slow down, take plenty of time and tend our educational gardens - a time for pruning, weeding, cultivating, fertilizing, and propagating.

Contents

It makes sense to have just enough technology,
just in time to get the job done. Just in time technology!

If we use an expensive tool infrequently, we usually rent it or we borrow from a neighbor. We learn to share, to take turns, and to be strategic.

Few families buy a computer for every person in the household even if they can afford it. Family members are expected to take turns. Plan ahead. Schedule your time.

The more expensive the tool and the more rapid its typical obsolescence, the more we might gain by strategic deployment. This holds true for schools, for businesses, and for families. Strategic deployment is well planned movement of equipment to maximize and optimize use.

Just in time technology is a notion whose time has come. Schools and businesses alike see the mounting financial burden of maintaining computer networks when software and hardware companies keep introducing upgrades and new versions that make rapid and frequent replacement of equipment imperative.

Just in case technology is a wasteful, even profligate approach to resource management that threatens to distort school and business budgets, seducing managers into delaying roofing repairs in order to maintain current, trendy digital equipment and networks.

1. Technology as Goal

For a decade now, schools have been urged by a procession of visionaries to equip all classrooms and all children with high powered, globally connected digital tools.

These promoters have put the cart squarely before the horse and sadly ahead of program development.

The tools became program.

"We are doing the computer."

"We are doing technology."

"We are doing the Internet."

"We are doing handhelds."

And state governments often poured urgency on this digital bandwagon by creating technology grants, curriculums, tests and standards as if technology were a content area rather than a set of tools.

School leaders found themselves pressured to network their schools before anyone had a very clear notion of how these tools would enhance student performance.

Not only did wiring classrooms become a priority - so did the purchase of many computers for each classroom as organizations such as the CEO Forum equated the level of curriculum integration with the number of computers per classroom.

Microsoft, Toshiba, Appl,e and other companies promoted "Any time, any where learning" and laptop schools as a form of educational Nirvana, rushing to equip many independent schools and pilot schools with a laptop for each student. One-to-one computing is all the rage.

We could label this kind of technology procurement and deployment the "just in case" technology model - a strategy that fills classrooms with equipment before clarifying purpose, value, or strategy. The discovery of purpose and learning strategies is evidently expected to occur some time following installation.

A decade after the creation of the first laptop schools, the evidence of value remains elusive and some program evaluations have proven quite disappointing. See data in Chapter 15 - "After Laptop."

This book argues for a dramatically different approach.

Just in Time Technology

II. Technology in Its Place

No computer or digital tool before its time!
   (or it **is** time)

Basic Beliefs

1. Avoid doing technology for the mere sake of technology.
2. Learn when to use classic tools and when to use new tools.
3. Know when to go unplugged.
4. Digital is not always best.
5. eCommerce is rarely a suitable focus for schooling.
6. People with little background or training in education rarely understand or appreciate the best strategies to improve classroom learning.
7. No computer or digital tool before its time!

In some quarters this **just in time** approach will be judged heresy, as the assumption for these folks is that more technology is always better.

But this book will argue that smart schools, students, and teachers learn to be discerning and strategic users of technologies. They share, take turns and move tools about to optimize returns on investments. When selecting a tool, they give full consideration to all technologies ranging from books and the human question to probes, digital cameras and PDAs. They also learn to unplug, turn off and tone down.

Smart schools are neither digitally obsessed nor technologically possessed. They have learned to say "No!" to distractions, silly toys and untested innovations that might reduce their focus on education's bottom line - the improvement of student learning.

III. JITT Strategies

This approach requires a good deal of cooperation, planning, and strategy. The following strategies often make the difference between success and disappointment:

Cultivating

Advance program and unit development is required so that scheduling and planning are possible. These units can then become required

Just in Time Technology

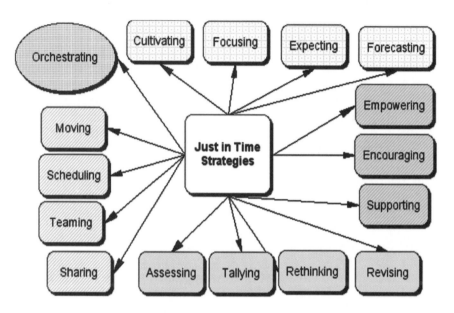

elements in each curriculum document with all teachers at a grade level expected to complete the unit. The creation of units reduces the uncertainties and the randomness of unplanned events.

Focusing

Develop technology rich learning units only in those disciplines and areas where it makes sense. Choose depth and quality over thinly distributed, superficial, and inconsequential uses.

Moving

Put the equipment where it will do the most good and move it frequently as needs shift. Lightweight laptop carts can visit 2-4 class-rooms a day without much difficulty. If one teacher only needs the equipment for a few periods, move it elsewhere during the "dead" periods. Beware of carts too heavy to move easily.

Sharing

Many learning tasks will flourish when pairs and trios of students share laptops or other tools. The same may be true for a team of teachers moving laptop carts about. One teacher may wish equipment in the morning but gladly give up afternoon access.

Scheduling

Who gets what when? Someone needs to take responsibility for deciding and communicating who will need how much equipment at which times. Haphazard allocation and movement is wasteful.

Forecasting

Someone needs to take responsibility for anticipating needs likely to arise during coming weeks and months.

Supporting

Making smart use of fewer computers requires support services to keep equipment in great shape and to help with the scheduling and movement.

Empowering

Equipment becomes an extension of the teachers' and the students' interests and volition - an invited guest meant to serve and strengthen. No more presumptions.

Encouraging

Those who take the risks of using new tools need to hear acknowledgment, recognition, etc. They also need emotional support when facing difficulties.

Expecting

No equipment sitting idle. Use it or lose it. Teach the curriculum. Make the units happen. Everybody.

Assessing

Gather data to see how various technology units are influencing student skills and performance. Shed, cull, weed, and eliminate those units and activities that make few discernible contributions.

Just in Time Technology

Tallying

Keep track of how much equipment is actually used. Set clear goals for usage and track real usage vs. that goal. Look at revising strategies to meet usage goals. No denial.

Teaming

Certain kinds of teaming can reduce the need for formalized scheduling and planning, as laptop carts may be shared across four teachers at a grade level if they have strong collaborative skills and inclinations.

Orchestrating

To avoid haphazard implementation, someone with clout (a principal?) needs to keep an eye on the system-wide use of equipment, noting its breakdowns, squeaking joints, lurches, delays, gasps, and stumbles. The larger the school and the more complex its culture, the more leadership is needed to keep the system humming smoothly and productively.

Rethinking

Based on experience, some strategies will prove more successful than others and the program must be reviewed with an eye toward shedding the poor strategies in favor of those with the best chances for success.

Revising

Periodic reviews and rethinking will lead to revision, synthesis, modification, and invention of new approaches and combinations of strategies.

IV. The Unwired Classroom

When computers first arrived in classrooms twenty years ago, they brought with them many headaches and challenges in the form of electrical cords, attachments, peripherals, and other entanglements. Sometimes it was hard to find room for them. Sometimes they felt like intruders.

In recent years, with the rush to network schools, the amount of wiring and cabling escalated dramatically in ways that often hampered the use of the equipment and led to restrictions and inflexibility. Connections often dictated placement. Movement was rare.

With the arrival of high performance wireless notebook computers, we stand at the beginning of a promising new phase. The wireless notebook, especially when delivered to classrooms in sufficient quantities, is likely to bring about a welcome shift in attitude and use by classroom teachers.

After several decades of limited progress toward widespread integrated use (as evidenced by reports such as **Technology Counts 1999** and the work of Hank Becker), wireless notebook computers may prove an ally in the effort to recruit the enthusiastic daily use of those teachers who have been hitherto reluctant, skeptical and late adopting when it comes to new technologies.

Wireless laptops can help to eliminate many of the barriers, obstacles, and inconveniences that have contributed to teacher reluctance in the past, but the focus of use must be on curriculum and learning.

If wireless notebooks are used for powerpointless activities, many teachers will wisely persist in resisting their use. We must offer standards-based learning experiences with these new tools if we wish to see broad-based use.

V. What are the advantages of wireless laptops under optimal conditions?

- Ease of Movement
- Relaxed Fit
- Strategic Deployment
- Flexibility
- Cleanliness
- Low Profile
- Convenience
- Simplicity
- Speed

In visiting schools to prepare this book, I frequently encountered situations that fell short of optimal. These challenges, frustrations and limitations of wireless computing are outlined at some length in Chapter Five - "Models of Movement." This chapter looks at the advantages of mobile computing when it is done effectively.

Ease of Movement

Laptops allow for easy movement within a classroom or across a school. Laptops also make possible the clustering and combining of enough units in each classroom to achieve critical mass - enough computers to do something worth doing.

Because desktop computers tended to be large and heavy, they required special furniture and were rarely moved around to where they might do the most good or be used most frequently.

Laptops are free to roam whereever students are learning. They require no special furniture. They can sit next to books and papers on a regular desk surface. They can fit on a lap. They can rest on a floor on a rug.

When one teacher is finished with a technology rich lesson and has little need for the laptops, they are simply loaded into a cart and rolled down the hall to someone who is ready and eager to begin a unit. No need to sit idle while non-computer tasks are taking place.

Ease of movement is likely to encourage more frequent daily use of each computer.

FNO began arguing for COWs (computers on wheels) and flotillas several years ago (see the March, 1998 issue, "The WIRED Classroom: Creating Technology Enhanced Student-Centered Learning Environments." http://fno.org/mar98/flotilla.html

But the use of flotillas and laptop carts was always somewhat frustrating and difficult because it was hard to move things around and manage the network connections as long as the computers required cables for electricity and network connections. Those who asked for movement were often told it was not a practical option.

Sadly, until now, the preferred model in most classrooms has been a thin and even distribution of 3, 4 or 5 computers per classroom, often bolted down at the back of the room in a fixed location on computer tables. This model provides far less student access than a wireless laptop cart with 15 units and proves frustrating to teachers who need critical mass to launch significant projects such as writing as process or WebQuests.

The math is simple.

If 25 elementary students need to spend 8 hours each on their writing project, they require 200 hours of computer contact time. Five computers provide 125 hours of contact time per week if used 5 hours each day during a 5 day week. It will take 2 weeks for these students to complete their one assignment if all five computers are used almost constantly.

But many teachers will not allow students to use computers all day long. While they are teaching math and other lessons, they may demand the full attention of the entire class. During this time, the computers sit idle.

In contrast, a laptop cart with 15 computers would provide the 200 hours in 5 mornings so two teachers could switch the cart from room to room and finish the writing in both classes that week. One teacher does writing in the morning while the other does writing in the afternoon. When they need to do other tasks, the computers leave the room and go next door where they will be used without pause for the rest of the day.

Thin distribution of resources has been a prime cause of the screensavers' disease - the lack of use by large numbers of teachers.

In many schools, computers were distributed to all classrooms regardless of the readiness or inclination of the receiving teachers to blend the new tools into daily lessons. Despite research by Becker and others indicating that use is heavily influenced by teachers' preferences, styles and readiness levels, all teachers and all classrooms in some buildings were treated equally. Ironically, the most equitable distribution of equipment to fixed locations may lead to a lack of real access and use. Becker (1999) found that constructivist teachers allowed almost three times as much use of computers as traditional teachers when they each had five computers in their rooms.

Frequent movement of equipment is likely to produce more true access and equity than thin distribution to fixed locations.

Relaxed Fit

Wireless notebook computers fit right into the classroom with little fuss or bother.

Unlike their desktop cousins, wireless laptops are quite small. They have a tiny "footprint" compared to desktop units. Because they take up very little space, they can sit down just about anyplace in a regular classroom without any special provisions being made.

With many teachers feeling skeptical and reluctant about using new technologies, a relaxed fit is a strong selling point. Laptops do not shift the room around or bring with them new pieces of furniture or other encumbrances.

Just in Time Technology

Strategic Deployment

Wireless notebook computers make it easy to put computing power where it will do the most good.

We should be applying new technologies to standards-based, curriculum-rich learning activities.

We should begin by asking what kinds of student learning we hope to promote. Those questions then logically lead to considerations of strategy and resources. Once we have a good sense of our purpose and the activities we plan to launch, we can begin to design a network that serves them well. Design should follow function.

In many schools, we see far too little consideration of movement. The prevailing strategy is to install and lock down all new computers. Yet this strategy is incredibly wasteful and inefficient.

Strategic Deployment involves a marriage of equipment and program. When the biology teacher is ready to launch a major study of the rain forest, we wheel a dozen networked computers into the classroom - enough resources to support genuine program integration.

Strategic Deployment takes us past tokenism and lip service to authentic engaged learning activities.

Moving computers where they are needed and wanted may allow a school to cut its hardware budget in half while slowing down the purchasing and replacement cycle. Instead of installing 2-3 computers per classroom that will be used (maybe) 15% of the time, the district cuts its order for 2000 computers down to 1000, invests heavily in professional development, and realizes 85% utilization by moving the equipment to where it will be welcomed (and used).

One week here. One week there. Movement spawns use!

Flexibility

Wireless notebook computers can be used in many different ways to support a lesson, with form following function and purpose.

In trying to recruit the enthusiastic participation of all teachers regardless of style and preference, the more flexible the delivery system the better.

A cart of laptops can be deployed across a room in many different ways, thereby maximizing the fit between the teacher's style and the way the lesson proceeds.

One teacher might prefer cooperative learning and teaming. No problem. Laptops lend themselves well to this type of lesson, as clusters of students gather around a single screen to consider and

10

analyze data.

A second teacher wants students working solo but facing the front of the room in rows. No problem. The students sign out a laptop and sit where they usually sit.

The key to this element is that life goes on as it normally would. The laptops allow the teacher to execute the lesson without having to move furniture or modify the existing norms and procedures.

Delivered to the classroom in sufficient numbers for short periods of time, the laptops provide enough information power so that a great lesson can be launched without any heroic scheduling strategies. This stands in direct contrast to rooms with 3, 4 or 5 computers that require teachers to perform balancing acts as less than half the class can be on the computers at a time.

With desktop computers, the teacher must be very flexible. The demands on the teacher to vary from normal routines is one of several complications that stand in the way of many teachers embracing new technologies.

The more comfortable and familiar we make the classroom experience with new technologies, the more likely we are to win enthusiastic use.

Cleanliness

Wireless notebook computers do not add mess, confusion and disorder to a classroom.

Finally we have computers that can sit on a desk with no more mess or bother than a textbook.

The typical desktop computer is not only large and bulky. With its many cords and cables, it can be downright messy. Those wires and cables either sprawl all over the place taking up lots of space on the desktop or they can be neatly hidden away behind and below specialized furniture that usually ends up virtually bolted against a wall.

The desktop computer ended up occupying too much space - dedicated space. Instead of entering the classroom like one more tool for daily use, it demanded special treatment. Ironically, this special treatment usually meant setting the equipment apart from the rest of the room. In many classrooms, the computers are off to one side.

While it may seem like a small thing, the neatness and simplicity of a wireless laptop gives it a huge advantage. The less trouble they cause, the more welcome they are likely to be. Teachers with a low tolerance for disorder and chaos will appreciate the simplicity and clean lines of these computers.

Just in Time Technology

Low Profile

Wireless notebook computers sit low, keep a low profile and allow teachers to keep an eye on things.

Those who have tried working with students in computer labs or classrooms filled with desktop computers will find the low profile of notebook computers a welcome change.

Walking around the room, it is easy to see students as they work with notebook computers. The screens are low down, close to the desk or table on which they sit. They tend to be no higher than a student's chest.

In contrast to a room of desktop computers where student heads are concealed behind monitors that are set up high, the teacher can keep an eye on things, judging from facial expressions who is on task, who is confused, and who needs some attention.

If the teacher wants to see screens while students are working, it is easy to configure the room that way. To move desktop units around in that way would probably be too much trouble.

One major barrier to widespread use of computers has been various control and classroom management issues that arise when students make use of such equipment. The tall profile and size of the desktop units contributed to the challenge of maintaining good attention and behavior during such lessons. Laptops return visual control to teachers in ways that make the classroom feel normal.

Convenience

Wireless notebook computers are easy and comfortable to use.
The American Heritage Dictionary (Third Edition) defines "convenience" as follows:

The quality of being suitable to one's comfort, purposes or needs.

The more a new tool matches teachers' purposes and needs, the more likely they are to welcome the tool's arrival. The more comfortably the new tool fits into the daily routines, the physical space, and the activities of the classroom, the more enthusiastically teachers will embrace the technology.

In one classroom, I watched a teacher using laptop computers for the very first time. After describing a research assignment, she told students to sign out a laptop and get started. In a matter of minutes, they were eagerly conducting the research.

Just in Time Technology

For a first time use, the high level of comfort and productivity was very impressive. Students picked up a laptop, went back to their groups, opened the equipment, and fired up their units without any trouble at all.

Because the equipment was so easy to use, the teacher could concentrate on teaching. No troubleshooting required.

In this example, the laptops made the teacher's job easier. They delivered information power to the desktop with no more bother than it would take to pass out a set of print dictionaries or encyclopedias.

In other classrooms and other schools, I heard many stories of difficulties and disappointments caused by an array of equipment issues such as brief battery life and network software demands. In those schools, the convenience of wireless was dramatically reduced and mobile frequently became immobile - wireless became wired.

In some districts, the laptop carts are so heavy that a single, strong adult will have great difficulty rolling the equipment any place. In one elementary school I visited, the carts stay in closets and students come down the hall for the laptops as they are needed. The laptop cart stays put.

Simplicity

Wireless notebook computers do not require much in the way of special effort or understanding.

Over the first two decades of their time in classrooms, computers have won a reputation for being a bit complicated and unreliable. The arrival of networks seemed to enhance that reputation in many schools and districts as the performance of a computer was now linked to the performance of the network.

"The network is down," signalled frustration and risk. Security concerns and risks added layers to desktops that seemed to keep users distant from the inner workings of the computers. IP addresses and TCP/IP added a level of mysticism to the Internet connected computer that could seem confounding to lay people.

Fortunately, we may be entering a new phase in districts that have passed through the early stages of networking. Network reliability has finally been achieved in many of these districts. Things generally work as you wish them to work. It is not such a big thing any more.

As described in the previous section, a cart wheels into a room, students pull out laptops and fire them up without any difficulty.

While this improved level of performance makes the use of both desktop units and laptops more simple and more appealing, the timing works especially well for wireless laptops, as they are arriving in classrooms at a time when a district can actually deliver "plug and play."

The chances for daily use are greatly enhanced by simplicity, comfort and reliability.

Speed

Wireless notebook computers can be deployed across a classroom or down a hallway in seconds if the cart is not too heavy.

Time is a major issue for teachers. Fortunately, laptops can be moved around quickly to match teacher wishes. A lightweight laptop cart is easily wheeled across the hallway at noon so two teachers can share the computers half a day each. There are no special connections, no TCP/IP changes, no fuss, and no bother. It is fast, easy, and comfortable when done correctly.

Movement within the classroom may also be fast.

Not so long ago I taught an afternoon professional development session at a laptop school that still had laptops requiring cables for a network connection as well as a power supply. It took more than 15 minutes to plug them into the wall! I wondered how it felt when students arrived in class and passed through the same hurdles.

This should no longer be a problem. Theoretically, the wireless notebook eliminates the "plug" part of "plug and play." The fuss, bother, and set up time associated with those wires, cables and network connections have been removed . Teachers can move right to the lesson. Students can focus swiftly on the learning. Unless, of course, the network software takes forever to log students onto the network. Unless, of course, there are battery limits and other issues.

VI. Other Issues

The advantages listed in this chapter are theoretical based on the availability of lightweight laptop carts, batteries that last an entire day, network software that is fast and reliable, and equipment that performs in a trustworthy manner.

Unfortunately, as is reported in some detail in Chapter Five - "Models of Movement," most schools I visited have found that there are many issues, problems and challenges when using these wireless technologies that have undermined their effectiveness and frustrated their users. This chapter posed the "best case" arguments for mobile computing. Other sections will detail some of the challenges and frustrations.

Riding the Curl of Innovation

Not so long ago, it was fashionable to speak about students surfing the Net. Schools rushed to connect classrooms to the Internet as if mere connectivity might work wonders. Many proponents of new technologies promised revolutionary shifts in the kinds of learning that would occur if schools bought the right equipment. They also predicted impressive gains in student performance - claims rarely substantiated by credible research findings.

But then the Internet and the dot com bubbles burst. Many ventures proved unworthy. Others turned into dot compost. Some schools awoke with empty hands and bankrupt business partners. Some digital emperors even paraded without clothes. At about this same time, the rush to wire classrooms was criticized by The Alliance for Children as a rush for "Fools Gold."

Given this recent history of speculation followed by skepticism, criticism and doubt, schools now face a menu of apparent opportunities seemingly laced with risks.

How can schools maximize a return on technology investments, backing mostly winners while avoiding losers? How can schools ride the curl of innovation without tumbling into heavy surf? How can they escape failure and a vicious undertow?

This chapter and the next together describe a strategic approach to the selection of innovative educational practices and tools - an approach designed to protect staff and students from "toolishness" - a fondness for tools that transcends purpose and utility. (**From Now On** - September, 2001 - http://fno.org/sept01/toolishness.html)

The goal is to improve schools without falling prey to bandwagons or train wrecks.

A Dozen Strategies for Making Discerning Choices

> Life's but a walking shadow, a poor player
> That struts and frets his hour upon the stage,
> And then is heard no more; it is a tale
> Told by an idiot, full of sound and fury,
> Signifying nothing.
>
> **Macbeth**, by William Shakespeare

How can we avoid what Shakespeare warned about some 500 years ago?

Discernment is the answer. We approach the adoption of new tools and practices with **discernment**.

discernment
1. The act or process of exhibiting keen insight and good judgment.
2. Keenness of insight and judgment.
   The American Heritage® Dictionary of the English Language: Fourth Edition. 2000.

Teachers and administrators may select from a dozen strategies to help make discerning use of new technologies. These strategies make it possible to sort through the noise of conflicting marketing claims to focus upon value, reliability, and authenticity.

1. Prospecting
2. Focusing
3. Challenging
4. Testing
6. Comparing
7. Remembering
8. Triangulating
9. Debunking
10. Deconstructing
11. Inventing and Evaluating Locally
12. Delaying

This chapter explores **prospecting** in depth. The next chapter describes the remaining eleven strategies.

Classic Prospecting for Oil - Convergence

We all have television-inspired images of aging prospectors with long beards who crisscrossed the desert with pack horses and little success. They may not have known enough science or applied enough strategy to the challenge.

For a photograph of a prospector, see the **History of the American West, 1860-1920:** Photographs from the Collection of the Denver Public Library.

http://gowest.coalliance.org/cgi-bin/imager?10060882+X-60882

Effective prospecting is a blend of art, science and skill, not simply a matter of wandering around with a divining rod in your two hands hoping to find the gold, water or oil below the surface.

Chevron shares the following information on its Web site describing the search for oil. (http://www.chevroncars.com/know/primer/)

"The Prospector's Primer from Chevron"

People have used petroleum products for nearly 5,000 years. The Babylonians caulked their ships with asphalt, and the ancient Chinese lit their imperial palaces with natural gas. For these early users, finding petroleum was a matter of guesswork and good luck. People simply looked for oil seeps and hoped the source was nearby.

Today, petroleum prospecting is considerably less random.

The goal is to find a **convergence** of the geologic elements necessary to form an oil or gas field. These elements include (1) a source rock to generate hydrocarbons, (2) a porous reservoir rock to hold them and (3) a structural trap to prevent fluids and gas from leaking away. Traps tend to exist in predictable places - for example, along faults and folds caused by movement of the Earth's crust or near subsurface salt domes.

Convergence and the Search for Value in the Technology Marketplace

Schools may also employ **convergence** as they scan the extensive menu of new tools and practices currently offered by eager vendors. The goal is to find convincing evidence of each of the following elements associated with value:

Element One - Designed for Learning?

Is this innovation solidly grounded in sound learning principles?

Those who sell tools and programs to schools do not always understand how classrooms, schools and learning take place. Rather than bother with what we know about effective learning practices, they skip the discussion by underlining how different the high technology classroom will be. This kind of sales approach is usually a danger signal. The wise school checks to make sure that the inventors of the new program understand the psychology and philosophy of learning in a way that matches the values of the school. Part of this checking is a review of the credentials and experience of the invention team responsible for the creation of the product. Does the team possess a strong track record of success within the educational workplace or are they outsiders with little knowledge of what works or doesn't work? Are they inclined to invent hybrids rather than focus on solid value?

Element Two - Tested and Refined?

Has this innovation been shaped and refined by careful field-testing and data collection?

In all too many cases, the product is rushed to market without much field-testing or refinement based on actual usage. When a school asks for data to guide decision-making, this request may be met with a blank stare or statements about leading edge technologies, vanguards, and pioneers. This is another danger signal. Product development should include careful testing and the collection of data. If the innovation has been field-tested without data and the vendors present nothing but glowing testimonials from heavily invested administrators, place some calls to find out how the rank and file feel about the innovation. Reliance upon testimonials rather than data should set off an alarm.

Element Three - Comfortable and Friendly?

Is this innovation user-friendly, inviting, and familiar to the rank and file?

If we seek broad-based acceptance of an innovation, we would hope that someone had taken the time to make it a comfortable rather than a threatening experience. Unfortunately, sometimes the designers of an innovation are out of touch with rank and file teachers and do their designing with a focus on the styles and preferences of early adopters. Look for evidence that someone has considered the needs and preferences of reluctants and late adopters as well as champions and early adopters. If the innovation is unfriendly in design, it will be hard for the majority to swallow and unlikely to thrive within the school.

Element Four - Effective in Winning Results?

Is this innovation capable of creating improvements in student performance?

The bottom line for schools is student performance and the likely impact of the innovation on the daily practice of the school. Schools should avoid doing technology for technology's sake and should only adopt innovations that have considerable promise to improve the reading, writing, and reasoning of students. There is little to be gained by bringing more toys and tools into schools that promote such things as powerpointlessness or cut-and-paste plagiarism. Does the vendor have hard evidence of gains substantiated by research conducted with some degree of reliability and validity?

Prospecting as Rigorous Questioning

The innovation prospector must dig below the surface, strip away the advertising claims, and figure out if there is gold in those hills. Schools that wish to make wise technology and program choices will heed the list of verbs associated with prospecting and searching by **Roget's Thesaurus**:

Inquiry: search (verb)

search, seek, look for
conduct a search, rummage, ransack, comb
scrabble, forage, fossick, root around
scour, clean out, turn over, rake over, pick over, turn out, turn inside out, rake through, rifle through, go through, search through, look into every nook and cranny
look or search high and low
search high heaven

sift through, winnow, explore every inch, go over with a fine-tooth comb
pry into, peer into, peep into, peek into
prospect, dowse, treasure-hunt, embark on a treasure hunt

from the 1996 **Microsoft Bookshelf** version of **Roget's Thesaurus** of English words and Copyright © 1962, 1982.

References

"Exploring for Oils." Schoolscience.Co.uk. http://www.schoolscience.co.uk/content/4/chemistry/findoils/index.html

"Fool's Gold: A Critical Look at Computers in Childhood." The Alliance for Childhood. October, 2000. http://www.allianceforchildhood.net/projects/computers/computers_reports.htm

"Foolishness is Toolishness**." From Now On - The Educational Technology Journal**, September, 2001. http://fno.org/sept01/toolishness.html

Ferdi Serim's article, "Gold into Straw: Alliance Report Misses Mark." at http://www.cosn.org/resources/113000.htm.

"The Prospector's Primer from Chevron" http://www.chevroncars.com/know/primer/

Two decades after the introduction of personal computers in the nation, with almost all schools wired for the Internet and nearly $6 billion spent annually on school technologies, the results get a failing grade.

> *Larry Cuban*
> *"System Crash."*
> **Boston Daily Globe**
> *November 25, 2001*

For decades, Larry Cuban has been noting the failures of new technologies to influence or improve the daily practice of teachers in their classrooms. Despite more than twenty years of fumbling efforts, we still tend to put the cart before the horse, believing foolishly that new laptops or handheld devices will teach Joan and Johnny how to read or grasp mathematical functions.

By now we should have learned that **toolishness** is foolishness. Showering fancy equipment and toys on classrooms without smart planning is unlikely to produce gains in student performance.

It is not that all new technologies must fail. Achievement of learning goals with new technologies requires a blend of powerful professional development, program development and **discernment**. There are many dumb ways to use the new tools that will create no changes worth noting. There are other smart uses that promise to sharpen the analysis, interpretation, inference and synthesis skills of students.

This chapter is about wise choices - **Smart Tech**!

When we approach the adoption of new tools and practices with **discernment**, gains in student performance are more likely to accrue.

As outlined in Chapter 2, teachers and administrators may combine and orchestrate a dozen strategies to achieve discerning use of new technologies. These strategies help sort through the noise of conflicting marketing claims to focus upon value, reliability, authenticity, and results.

Smart Technology 2

1. Prospecting
Looking for the right combination of promising program elements and indicators.

This strategy was fully outlined in the previous chapter.

2. Focusing
Keeping an eye on prime philosophical commitments and program purposes.

This effort is not about networking schools just to be networked. It is not about installing the latest technologies and digital toys just to stay ahead of neighboring districts or schools. We must not embrace technology for technology's sake. It is neither powerpointlessness nor Tomfoolery nor mere handholding (as in handheld devices). We should only acquire new technologies that will improve student performance on learning tasks that match state curriculum standards or address important local learning goals.

Smart districts create brief, 4-5 paragraph summaries of the learning goals to be addressed with new tools. These clear statements are drawn from the staff, refined, carefully discussed and then widely distributed once final drafts are approved. Walk into any classroom and ask the teacher what the district is hoping to achieve, and she or he will be able to answer succinctly.

"We hope to improve the information literacy skills of all our students by challenging them with problems drawn from the curriculum. We emphasize analysis, inference, interpretation, and synthesis - just like the state standards."

Such belief statements help to screen out unworthy technology programs, products, and activities. If programs do not deliver the kind of student learning chosen by the district, they do not qualify for purchase. Virus protection!

For more on this strategy, consult "First Things First" in the November/December, 2000 issue of **From Now On** at http://fno.org/nov00/f1.html.

3. Challenging
Demanding evidence, data, results, and substantive theoretical underpinnings. Considering the risks, the costs, and the dangers.

We are often assured by vendors and program cheerleaders that new tools and products will shift classroom practice and results dramatically. But where's the beef? These claims and promises are rarely substantiated with program data gathered in a credible fashion. In most cases, the data is what might be called "testimonial research" - meaning that we are shown a bunch of rave reviews by pioneering teachers and administrators who swear that the classroom, the school and the district were transformed by the innovation.

"But what happened to the reading and writing scores of students?"

In those cases with numerical data showing student progress, the design of the experiment is often seriously flawed, allowing volunteer, pioneering teachers to teach the experimental classes, for example, while non volunteer teachers are left as control groups. The resulting "progress" may be heavily tinged by the special qualities of the volunteer teachers as well as what is known as the Hawthorne Effect - the tendency for any group getting special attention to improve performance because they are being watched,

For decades, consumers have turned to **Consumer Reports** to compare the gas mileage, safety, and repair records of various automobiles. The same kind of data gathering makes sense when it comes to technology shopping.

But the data should include more than evidence of success. A smart technology planning team looks at the dark side of any innovation and tries to discover what might possibly go wrong. Technology proponents and vendors will often hype the innovation and emphasize nothing but the positive. Without a full understanding of the risks and challenges likely to accompany the product or program, the chances of making a successful launch are reduced significantly. See "The Post Installation Action Plan" in the November/December, 2001 issue of **From Now On** at http://fno.org/nov01/postinstallationplan.html

4. Testing
Setting up small, low-risk pilot programs and reviewing the results of others' pilot tests.

The fashion these days is to plunge into major initiatives without running pilot programs to assess their worth and then determine how to adjust the programs to work well under local conditions.

This elimination of conservative and cautious change strategies is a side product of the boom times of the 1990s as industries and schools joined together to build electronic highways that seemed to promise wondrous futures for us all. Sadly, the dot com bubble burst and the extravagant investments in infrastructure have not produced the expected profits and benefits in either world.

5. Investigating

Looking past the surface claims to find out what really happens when the tools and practices are installed and implemented. Finding prior innovators to learn the "true story" of what happened.

Which implementation models are most likely to produce learning gains?

So as to avoid repeating the mistakes of early adopters, if an innovation has already been field tested by other schools, a wise team devotes substantial time and attention to considering what others have learned. This learning requires real tact and skill because many early adopters may promote the innovation without sharing stories of frustration and disappointment.

The trick is to find innovators who can be trusted to share the trials and the tribulations as well as the triumphs. This search for verity often requires some probing past leadership levels to see what rank-and-file teachers report about the innovation experience.

The important words are "likely to produce learning gains." This is not about flash or excitement and fireworks. We are looking for solid results substantiated by concrete data. Testimonials are abundant but rarely reliable.

When early adopters claim glorious results, it pays to ask, "How do you know? What evidence do you have?"

6. Comparing

Examining the full range of choices (vendors and models, for example) within a category along with alternatives that are substantially different.

In some cases, options may narrow too quickly. What seem to be special opportunities sweep the district along a path that does not allow for comparison shopping. Perhaps a vendor offers a "deal" that means the district will be first in the region. This kind of program glory may seduce leaders into premature commitment. It may even feel comforting for some time to focus energy on a narrow range of program strategies, but wisdom calls for a more deliberate approach that weighs the pros and cons of various approaches, avoiding traps associated with infatuation.

7. Remembering

Reviewing past experiences with innovations (and vendors) that promised similar results and changes.

Unfortunately, memory can fail - especially when it comes to recalling missteps. Denial is a serious liability when trying to learn from one's previous mistakes. Institutional memory loss allows a pattern of blunders to persist.

The serious study of organizational history is an essential safeguard against perpetual stumbling.

> *Those who cannot remember the past are condemned to repeat it.*

George Santayana

To some extent, the process of implementing innovations requires a trial-and-error adjustment of strategies in line with experience. Lessons from the past should prove illuminating and instructive when launching new efforts, but in many cases, the leaders of the earlier efforts may have moved on, taking history with them, or the organization may be inclined to admire the emperor's old clothes as well as the new ones.

A healthy organization heeds the lessons both of research and history. For more on this approach to leadership, consult "Beware the Shallow Waters! The Dangers of Ignoring History and the Research on Change in Schools," in the June, 1999 issue of **From Now On** at http://fno.org/jun99/teach.html.

8.Triangulating

Developing multiple, independent, potentially conflicting sources to support the evaluation process.

In the search for verity, we compare and contrast testimonials. We seek out the naysayers in order to balance and assess the claims of the cheerleaders and the promoters.

When the principal or superintendent claims heavy and effective use of the innovation, we ask for data.

"How do you know that?"

If the teachers report 30% utilization weekly of curriculum-rich, technology-enhanced lessons, what do their students report? Do the numbers match?

If we simply take notes when sitting at technology conferences, we become the blind following the blind.

Informed skepticism requires some detective work and some off-the-record conversations. Much of this negative content would not normally find its way to conferences or into publications inclined to trumpet the wonders of technologies (and advertisers).

School leaders need workshop sessions and articles that share openly the obstacles and challenges one is likely to encounter. In those same sessions, innovators offer antidotes and strategies to avoid those pitfalls.

9. Debunking

Stripping off the hype, the marketing claims, the myths, and all excessive promises to consider the prospects for success rationally and analytically.

Technology cheerleaders often create pressures for change and the purchase of more equipment by suggesting that the failure to keep up with the toys of one's neighbors will lead to severe deprivation - a future marked by unemployment in some digital (dot com?) work-place. The "Digital Divide" sometimes seems more like a marketing ploy than a reality, as vendors wring their hands over the gap (in purchasing) between rich and poor districts. The vendors' concern seems disingenuous at best, since recent Market Data Retrieval reports (cited in Cuban's article) show little effective use of networked computers by most teachers.

The real question - mentioned earlier - is whether this innovation will produce learning gains. The best measure of this goal would be educational assessments aimed at the analysis, interpretation, inference, and synthesis skills required by tough state or provincial curriculum standards.

Technology cheerleaders are fond of separate technology standards, assessments and programs, enshrining such dubious skills as powerpointlessness and handholding as major program objectives.

Educators should focus on teaching students to read for meaning in books and on screens. Reading e-books on a handheld device is a trivial goal. Interpreting a passage from **Macbeth** is challenging regardless of whether the words appear on paper or a screen. Providing handheld **CliffNotes** is unlikely to promote original thought or develop interpretative skills.

When new technologies make it easier to slide by on the thinking of others, they do this generation of students a disservice - a point that some companies seem to miss as they announce the availability of handheld **CliffNotes** with pride.

Promoters of new technologies offer packages of fabulous new century skills that suggest that being modern requires being equipped. Ironically, many of the skills in their package are not at all new. Many can be taught without digital devices. Their not having been taught in the past was not because we had the wrong tools. The explanation for failure to teach those skills is rooted in social and organizational patterns that are deeply set in our culture.

Hank Becker (1999) reports that some 70 per cent of the teachers in his national study are quite traditional in their approaches to class-room activities. There is no evidence that these teaching styles or strategies change when computers are placed in the classroom. To the contrary, Becker found they let students use the equipment one third as often as their more constructivist colleagues.

Teachers since Socrates have struggled to develop the thinking, problem solving and questioning skills of their students, but now we are told that an educated citizenry must be wired and digitally savvy. The implication is that the new technologies will promote a new, more powerful literacy. Yet we see little attention to emerging problems associated with such tools - the powerpointessness, cut-and-paste thinking, new plagiarism, info-glut and mentalsoftness reported by many teachers and commentators who see the Internet and digital resources as a mixed blessing.

The real question is whether this innovation will produce learning gains.

10. Deconstructing

Breaking the innovation into its component parts to see how well they fit together, how they are meant to work and where the vulnerabilities may lie.

How is this program supposed to work?

Packaged programs imported from outside a school district rarely succeed unless customized to match local conditions and special needs.

The process of deconstructing a program is a bit like taking apart a lawn mower or automobile engine back in the days when there were few electronic parts. By disassembling the motor or program into its parts, program managers may consider modifications and adjustments.

A district might identify a successful writing program that employs Inspiration™ and wireless laptops, for example, to teach students writing as process within the guidelines and expectations of state standards. Upon close examination and deconstruction, the district team determines that the program needs bolstering at the idea generating and the editing phases. Because the district has already invested heavily in an approach to editing and revision called "The Six Traits Approach to Writing," the team protects that investment by adding it as a major program component. They then go in search of idea generation strategies to enhance the model.

In many cases, innovations proceed to installation as if such tinkering and adjustment is unnecessary and burdensome, but a failure to modify a program may undermine past efforts like the one mentioned above and may also limit the potential of the new program to shift daily practice.

11. Inventing and Evaluating Locally

Engaging local staff members in the development and testing of innovations so they have firsthand knowledge of what works and what does not work, thereby reducing dependence on and vulnerability to outside promoters.

Homegrown innovations may stand a reasonable chance of taking root, especially if the district nurtures the development of human resources devoted to the invention process. This process requires cultivation, encouragement, and investment.

See " Creating the Vanguard: Identifying, Grooming and Rewarding the Champions" at http://staffdevelop.org/cadre.html.

Metaphors drawn from the world of gardening may provide illuminating insights for those planning innovations. If we are doing our jobs well, we will be raising seedlings in hothouses, testing them, clearing areas, weeding, cultivating the soil, fertilizing, planting, weeding, pruning, thinning, and weeding some more.

We are careful to plant where conditions are right. Unfortunately, some school planners leave out some of the most important steps in the planning (or gardening) process, installing a network without cultivating the soil - without investing in program development or professional development. They may plant desktop units where there is little willingness to use them, the equivalent of planting sun-hungry plants in heavily shaded areas.

Most of these missteps occur because these planners do not view the change process as organic. They focus on wires, cables, and equipment. They neglect the human and organizational elements that are basic to a thriving and robust effort.

But invention is not enough alone. Unless informed and guided by astute data collection to assess which aspects are proving effective, the innovation can turn into a new suit of clothes for whatever emperors are shopping.

12. Delaying

Slowing down the purchase and installation process so that schools can learn from the mistakes of others - avoiding the bleeding edge of change.

> *Slow down, you move too fast.*
> *You got to make the morning last.*

> Paul Simon

When it comes to networking schools and classrooms or adopting new technologies, some leaders tend to rush things. Just as folk wisdom argues that "Haste makes waste," research on change in schools warns against riding change mobiles through avalanche territory (Fullan, 1991).

Even though the shortest distance is usually a straight line, most road and railway builders know better than to head straight up a mountain without providing plenty of curves. They rely on switchbacks to keep the angle of ascent reasonable. They know engines have their limits.

Schools should follow this example. They should make the strategic pacing of change a priority if they hope to see a real (and beneficial) shift in classroom practice.

See " Pacing Change," an article that originally appeared in the September, 2000 issue of **Classroom Connect**. http://fno.org/nov00/pacing.html.

There are many advantages to delay when it comes to new technologies. School districts that come to the game late may find that slow and discerning decision making means they leapfrogged over some of the silly and misguided efforts of their neighbors.

Sometimes it turns out that we end up ahead of the game by letting other fools rush in.

References

Becker, Henry. 1999. "Internet Use by Teachers" Web site at University of California Irvine http://www.crito.uci.edu/TLC/FINDINGS/internet-use/startpage.htm.

Cuban, Larry. November 25, 2001 "System Crash." **Boston Daily Globe.**

Culham, Ruth and Spandel, Vicki "The Student Friendly Guide to Working with Traits" at http://www.nwrel.org/eval/toolkit/traits/.

Fullan, Michael. 1991. **The New Meaning of Educational Change**. Teachers College Press.

Chapter 4 - Post Deployment Action Plan

Worthy, frequent use of new equipment requires careful consideration of program strategies as well as professional development to cultivate the faculty's readiness. There are also many organizational strategies that may increase effective use. Schools with strong post installation plans are more likely to win a healthy return on their technology investments.

A post installation plan will address issues and questions like the ones below. Laptop schools, handheld devices and rolling carts of wireless laptops are used for illustration purposes, but the same kinds of questions might be applied to the introduction of any new tools or strategies.

1. Best Models

Which models are most likely to produce learning gains?

So as to avoid repeating the mistakes of early adopters, if an innovation has already been field-tested by other schools, a wise team devotes much time and attention to considering what others have learned. This learning requires tact and skill because many early adopters may promote the innovation without sharing stories of frustration and disappointment.

The key words in the question above are "likely to produce learning gains." This is not about flash or excitement and fireworks. We are looking for solid results substantiated by concrete data. Testimonials are abundant but rarely reliable.

If you check the closets of some schools, you are likely to find an ample supply of dusty bandwagons from the past whose glories were trumpeted by early adopters during the first two years of installation.

Discovering what has really happened at a laptop school (or a school with laptop carts or handheld devices) is sometimes a struggle against political forces that are heavily invested in the appearance of success and the stifling of dissent. Doubt, skepticism and honest appraisals may be characterized as heresy rather than wisdom.

The purpose of this team inquiry is to identify those strategies most likely to succeed in a new setting while weeding out strategies that prove to be of questionable value or significant risk.

The outside team must interview and probe beyond the front lines of corporate and educational marketing, speaking with those who work within the school. Rank and file critics are too often dismissed as slackers or heretics. In fact, their skepticism can sometimes be the voice of experience and wisdom.

How do you know?

When early adopters claim glorious results, it pays to ask, "How do you know? What evidence do you have?"

Some promoters and cheerleaders tend to answer with exuberance unsupported by data.

Razzle-dazzle

"We can see it in the students' eyes and their enthusiasm."
 "The teachers say the kids love it."
"We can't get students off the laptops."

The same could be said about arcade and video games.

The question deserves repetition, "What evidence do you have that this program produces learning gains?"

Some vendors seem incapable of grasping or respecting educational values in any real sense. How could any company forging partnerships with schools to improve student results also brag about supplying **CliffNotes** to handheld devices?

What's wrong with this picture?

Is the sudden infusion of **CliffNotes** via handheld devices really an improvement in high school English? What will the English teachers say? Should we applaud the arrival of handheld **CliffNotes**?

Why bother reading the real Hamlet or Moby Dick when you might fake your way through class with your handheld device?

Palm recently announced with apparent pride a partnership to provide access to **CliffNotes**.

A Post Deployment Action Plan

Quoting from an October 31, 2001 Palm news release:

Palm is building a thriving ebook business for consumers, mobile professionals, educators and students," said Jeff Strobel, director of Palm Digital Media.

CliffNotes have enjoyed immense popularity among students worldwide for more than 40 years, and Palm Reader is an ideal reading platform for learning and teaching, with features that include auto-scrolling, bookmarks, the ability to annotate, large fonts and several viewing options.

We expect students and teachers to gravitate quickly to ebooks as more academic books and materials become available in this format.

In August of 2001, the **New York Times** disputed these rosy predictions by ebook vendors and provided data to show that sales of ebooks have fallen far short of vendor claims and predictions.

Forecasts of an ebook Era Were, It Seems, Premature
By David D. Kirkpatrick (from the August 28, 2001 New York Times)

Laurence Kirshbaum, chairman of the books division of AOL Time Warner (news/quote), pledged to lead the charge: "We want to see electronic publishing blow the covers off of books." Andersen Consulting had recently estimated that by 2005 digital books could account for 10 percent of all book sales.

A year later, however, the main advantage of electronic books appears to be that they gather no dust. Almost no one is buying. Publishers and online bookstores say only the very few best-selling electronic editions have sold more than a thousand copies, and most sell far fewer.

It is worth noting the types of ebooks Palm has been selling. Quoting from an October 31 Palm news release . . .

Palm Digital Media Top 10 Best-selling ebooks October 2001

"Black House" by Stephen King and Peter Straub (Fantasy)

A Post Deployment Action Plan

"The Talisman" by Stephen King and Peter Straub (Fantasy)

"Straight From the Gut" by Jack Welch and John A. Byrne (Business)

"Rift In Time" by Michael Phillips (Fiction)

"K-PAX" by Gene Brewer (Science Fiction)

"Star Trek: S. C. E. #8: Invincible: Book Two" by David Mack and Keith R. A. DeCandido (Star Trek)

"The Ice Limit" by Douglas Preston and Lincoln Child (Espionage & Thrillers)

"Blood and Gold" by Anne Rice (Horror)

"The Coming Anarchy" by Robert D. Kaplan (Non-fiction)

"By the Rivers of Babylon" by Nelson DeMille (Espionage & Thrillers)

What would the English teachers say?

Jaws?

School leaders and planners need to be cautious about business partners who suffer from excess inventory, excess capacity and disappointing sales and profits. These conditions can induce hungry behaviors and intense promotion. Sometimes desperation may inspire giveaway programs that help reduce excess inventories while distorting educational agendas by flooding classrooms with products the schools might not have chosen if they were costly.

When philanthropy is thinly veiled marketing, schools should view such programs as Trojan horses. Some have jokingly warned, "Beware of Geeks bearing gifts!"

What if someone built a highway and hardly anybody used it?

In his article, "In Excess," published in the November, 2001 issue of **Attaché**, the U.S. Airways flight magazine, Daniel Gross, a fellow at the New America Foundation comments . . .

From Wall Street to Silicon Valley, investors and technology gurus are lamenting the excesses in the fiber-optic-cable market. Fueled by visions of an infinitely expanding need for Internet pipelines that would carry voice and data traffic, pioneering entrepreneurs and established companies went on a building binge in the 1990s.
(section skipped)
But something happened along the way to cyber-Nirvana. Neither the expected traffic nor the projected revenues materialized as

predicted. And so, on any given day, just 5 per cent of the fiber optic cable is pulsing with information.

In addition to requesting evidence, how does a team protect against false or exaggerated claims?

One tactic some researchers employ is "triangulating" -the tactic of finding several independent sources to play against each other.

The outside team is looking for models that have real promise - models grounded in classroom practice and classroom results rather than rhetoric, advertising and dot com fantasies.

What are the key elements of each model?

What are the prime strategies?

In the search for verity, we compare and contrast testimonials.

When the headmaster claims heavy and frequent use, we ask for data.

If the teachers report 30% utilization weekly of curriculum-rich, technology-enhanced lessons, what do their students report? Do the numbers match?

If we simply take notes when sitting at technology conferences, we might become the blind following the blind.

Seeing is believing.

School visits are an essential. This is no time for distance learning. "Take it on faith? No chance!" The district must provide a healthy travel budget so the team can visit far and wide. Chances are, the best models will be dispersed across the country and will require substantial travel.

But teams should be wary of schools that offer guided tours with promotional overtones. The best visits tend to be to schools whose leaders openly discuss mixed results and seem to value verity.

Cheerleaders, trend-setters, enthusiasts, and owners of stock are sometimes guilty of toolishness - a fondness for tools that transcends purpose and utility . . . as when folks grab a hammer to paint a flower just because they like hammers or because hammers are trendy or when they allow a computer to speak for them to an audience instead of telling their stories with a natural voice or when people turn to search engines to find truths more likely to reside in books or their own hearts.

Toolishness is closely associated with other terms such as foolishness, powerointlessness, microoftness, mentalsoftness, disneyfication, edutainment and infotainment.

For more on avoiding or noticing this trap, see the September,

A Post Deployment Action Plan

2001 issue of **FNO** at http://fno.org/sept01/toolishness.html

2. Broad-based Usage

Which school strategies (such as scheduling and unit building) will lead to usage by all teachers and prevent domination by a few?
Without taking strong measures to facilitate, encourage and structure the broad-based use of a new technology, there is an ever present danger that a minority of the teachers (the early adopters) will corral the equipment so that reluctants and late adopters might claim that they cannot even win access. This phenomenon often leads to a kind of "separate peace" as enthusiasts taste abundance while reluctants go about business as usual.
The antidote? The creation of standards-based unit plans at each grade level provides practical, well designed lessons that should win the respect of all teachers. Once these units are embedded in the curriculum, the school designs a schedule for the broad-based use of new technologies. For more on this approach, read the November, 1999 issue of **FNO** at http://fno.org/nov99/standards.html. "Teaching to the Standards."

3. Curriculum-Rich Units

What are some of the best choices a principal can make to support media specialists and teachers implementing curriculum rich units with rolling carts?
Drop shipments of new tools rarely promote significant use. Once they are unboxed and unwrapped (a process that can extend over several months in some districts), the real work begins. Given already full schedules and commitments, few staff members have the time and resources to figure out, test, and modify effective program strategies with the new tools.
Principals may promote innovation by clearing away obstacles, rearranging resources and establishing an organizational culture supportive of growth and risk taking.
Growth rarely occurs in schools caught in a survival mode. To move away from survival to healthy growth, principals may turn to online resources such as the **Journal of Staff Development** and its library of articles, many of which suggest strategies to reorganize schools and schedules to free up additional time for adult learning.
The principal should be orchestrating professional development

offerings to dovetail with the introduction of new tools. Often the new tools prompt a 2-3 hour introductory workshop . . . and then silence. The support should be ongoing and well spaced through the school year - just in time learning as opposed to just in case learning.

In order to overcome obstacles and frustrations, part of the learning should focus on classroom management strategies rather than mere equipment operating.

The principal should be sheltering staff from excessive demands, from frivolous program experiments and from topsy-turvy surprises and disruptions.

4. Evidence of Success

If the new program is successful, what data could a principal share with the superintendent and other supervisors as evidence of success?

The school can gather evidence that the new tools are being used frequently and according to plan. Furthermore, the staff can indicate that such use is universal. It is a simple matter of scheduling and tracking use. More important than frequency is the alignment of use with the local curriculum, state curriculum standards, and the school's plan to produce student learning results.

Alongside the schedule recording number of hours, there is a place to record activity codes from the list agreed upon by the staff.

Activity A - Text Literacy Challenges (analysis, inference, and interpretation ---> synthesis)
Activity B - Numerical Literacy Challenges ((analysis, inference, and interpretation ---> synthesis)
Activity C - Visual Literacy Challenges ((analysis, inferenc,e and interpretation ---> synthesis)
Activity D - Problem-Solving Challenges
Activity E - Decision-Making Challenges
etc.

Beyond recording the time and the ways tools are used, schools should also measure changes in student skills, performance and attitudes.

When technologies are not available in sufficient numbers and depth to reach all classrooms, the school should make a sincere effort to create control groups that are quite comparable, avoiding traps like the assignment of early adopters to the experimental groups.

A Post Deployment Action Plan

Do the state test results of students participating in these pilot programs surpass the results of those without the new tools? Or is the reverse true? Can we see any differences at all?

5. Expectations

How can clear expectations help to stimulate effective use?

In many schools, new equipment is purchased and rolled into buildings without the staff agreeing in advance upon how, when and why the technology might be used. This lack of definition and clarity allows each individual to do as she or he pleases and leaves the success of the innovation to happenstance.

"Hear no evil, see no evil, speak no evil."

Without assessment and clear expectations, participants are free to go about their business as usual. Bandwagons have passed through schools this way for decades, with little discussion or consideration of changes in student performance.

The remedy? The staff buildscurriculum-rich units involving the effective use of the new tools. If the fifth grade science curriculum calls for every student to do a stream study using handheld devices to analyze data, then all teachers are expected to schedule for the use of equipment and all teachers are expected to do the unit. The principal works with all teachers to make certain they have the resources required to meet the expectations clearly expressed in the curriculum guide.

6. Support Strategies

What informal support strategies and resources can be launched and sustained to extend the benefits of introductory professional development sessions?

Regrettably, most teachers must work through innovations in relative isolation from others. This isolation increases the sense of risk while enlarging the enormity of what Michael Fullan calls the "daily press" of classroom teaching. These pressures and the anxieties occasioned by disruptive new programs create a sense of "battle fatigue" in some schools. This fatigue can make the introduction of new learning tools and strategies seem daunting. Classroom routines become an important survival strategy under these conditions, and those routines are not readily exchanged for new ones.

Encouraging partnerships can make a big difference. When combined with liberal doses of time for meeting, discussing, plannin,g and launching, partnerships can give teachers the boost in spirit they need along with the skills required to attain good results.

In all too many schools, pressures for change are heaped on top of already stressed performers without considering the impact these pressures might have on the spirit of the staff as well as their inclination to welcome and embrace the innovation.

Many "change agents" project visions and lofty goals without taking care of the needs and feelings of those charged with carrying out the new marching orders.

The focus in many schools is the provision of formal training in software. While everyone needs software skills, there is a tendency to teach lots of skills out of context. Seeing no practical use for many of these skills and having no occasion to practice them, many teachers find that skills slide away like runoff after a storm.

According to research by Bruce Joyce, innovation is more likely to stick when teams of teachers learn and practice new strategies together over a substantial period of time.

In order for this kind of informal teaming to thrive, most schools must commit to organizational change.

- **Resources** - Funds must be dedicated to adult learning, providing time, opportunity, and staffing to keep teams moving forward.
- **Attitudes** - It takes some courage to work in teams on new behaviors that might cause discomfort or embarrassment. Positive attitudes usually require cultivation. The leader or leadership team might begin with a survey of attitudes to determine which deserve attention.
- **Norms** - The basic belief system, the rules, and the expectations of the organization can have a profound effect on any innovation's chances. As Peter Senge explains in considerable depth, several levels of beliefs may require attention. Sometimes the substrata go unrecognized and untreated even though they may block, weaken, and frustrate particular change efforts.
- **Procedures** - Within this category would fall organizational habits of being and acting - the types of meetings, the ways of sharing concerns, the kinds of messaging and communicating, etc. In some schools, these procedures can lead to a "closed system" - one that is impervious.
- **Structures** - Time schedules and other organizational structures can ease the passage of innovations or frustrate and hamper them. Do the third grade teachers ever get a common planning period?
- **Goals** - What are the real purposes of the organization? Some

schools are focused on the adults that work there. Others are fervently committed to engaging and challenging all students. These goals must be congruent with the innovation of the innovation is going to take root.

7. Best Practices

Which school districts have already discovered good ways to move past installation to challenging, standards-based learning with frequent and broadly distributed use?

The answers to this question depend upon the kinds of visits described earlier in this article. During the visit, the team tries to focus upon specific strategies that have helped to make lift-off a reality.

Chapter 5 - Matters of Movement

Mobile Computing? Easier said than done!

Even though wireless equipment can be moved about and shared, far too many schools are finding movement difficult. In some of the schools I visited in preparation for this book, the laptop carts are so heavy, they are never moved.

These frustrating aspects of movement are rarely mentioned in promotional material, but my visits to school sites uncovered many stories of surprises, difficulties, and challenges. While schools had managed to address most of these difficulties, the flexibility and spontaneity of movement hoped for prior to installation was frequently compromised in order to sustain reliability. Even such stories of success were laced with tales of struggles to overcome obstacles.

The Wired (Wireless) Laptop

Depending upon the brand of laptop purchased, short battery life forced many schools to plug laptops into power sources. In one high school, the prime candidate for laptop carts was the science department with its heavily wired lab areas. With network drops and electrical outlets at every lab station, the "wireless" laptops were fully wired and free from many of the troubles reported when laptops were truly wireless.

The science department was fully engaged in moving, sharing, and deploying the laptops because they could be counted upon to arrive in great working condition just in time. They also had little need for laptops to move around the room, since students would be spending most of their time in fixed lab positions, but if a teacher did want students to perform some computer task offline in the regular part of the classroom, the laptops could be disconnected from the power source for brief periods without depleting their usefulness for the next class.

Matters of Movement

An English or social studies teacher looking for much more movement and flexibility within a regular classroom would not find power and network outlets to support the kind of wired use enjoyed by the science teachers. They would consider the need to plug laptops into power strips a major limiting factor. In that school, the science department was the only department making frequent use of laptops.

This school found that truly free-ranging wireless laptops often suffered performance slow downs due to demanding network software. In one program making use of 50 truly wireless laptops residing in a double classroom, the network software had to be removed from the laptops. The software's "overhead" had placed a heavy load on the wireless operation within that space.

"They call it Mobile Computing, but they didn't tell us the cart would be too heavy for any mortal teacher to push down the hall and we'd need to rent an 800 pound gorilla!"

Matters of Movement

The Immobile Mobile

The school mentioned earlier found that the truly wireless laptops worked best when they were kept in one location all year. They kept 50 laptops in one double classroom space that allowed network staff to stabilize the system and iron out problems that frequently cropped up when carts were moved up and down hallways.

This school was excited about the potential of wireless mobile computing as a delivery system, but labs of desktop computers remained the primary access for classes and students. The prospect of replacing some of those labs in the next few years with laptop carts was attractive, but movement in that direction had slowed down as the mobility and performance of wireless laptops proved more quirky than expected.

Given a choice between a somewhat risky delivery system (mobile wireless carts) and a more stable one (labs), teachers were still inclined toward the predictability of desktop units plugged into infrastructure.

Even when everything functions properly, technology-rich activities are generally seen as challenging by teachers. They have little tolerance for network systems and computers that crash or stumble. The greater the stress associated with the activity, the smaller the cadre of teachers willing to give such activities a try.

In many of the buildings I visited, truly wireless use of the laptops was rare, as compromises were seen as necessary to deliver good performance.

In one elementary school, the laptop carts were too heavy and the stairways too numerous to allow easy movement, so they were permanently parked in closets with copy machines, and students had learned to walk down the hallways to pick up the laptops needed for that day's lessons. Far from disappointed with this system, the school had come to embrace it as quite practical and as surprisingly evenhanded, as they found that laptops circulated with more speed in that way.

Cartage Fees

Some schools allow teachers to sign out carts for days and weeks at a time, while others encourage more frequent movement, dividing up days into thirds at the elementary level, for example. In such schools, teachers are expected to roll the laptop cart back and forth so that other classrooms have a chance to use the equipment.

Matters of Movement

This scheduled movement and sharing maximizes daily use, eliminates down time and reaches the most students at the least cost, but the strategy is costly in human terms, as the stress and effort required to roll carts back and forth is mentioned by many teachers as a real problem.

Sharing a cart within a particular department can alleviate many of these "cartage fees" if all of those classrooms are clustered in the same part of the building. Likewise, several middle school teams in one wing of a building might share a laptop cart for their area.

Some schools create mini labs of laptops in common public hallway spaces within view of classrooms and make students move to the computers.

Security and Procedures

Concerned about batterry lfe, damage, and theft, many schools have developed quite demanding procedures that can reduce the attractiveness of mobile computing. While such procedures seem necessary, they can frustrate busy classroom teachers.

One district found that students were generally not reliable when plugging used laptops back into the cart's recharging system. Those uncharged units might then be unavailable for several classes and would cause unpleasant surprises. To guarantee complete charging, the district strengthened its expectation that teachers would personally plug all units back into the cart.

Many schools have tightened procedures in ways that increase the load on the teacher who hopes to use the carts. Well intended though these procedures may be, they may become barriers to frequent, comfortable use. The more burdensome the regulations, the smaller the likelihood that this system will be embraced broadly by staff, many of whom may have little time or appetite for heroics.

If teachers can inspire students to act responsibly and minimize the need for such procedures, broad use of the equipment is more likely. Likewise, if vendors can create products less susceptible to human error, more teachers and schools will welcome mobile computing.

Plug and Play

Every school I visited had come to view mobile computing as a delivery system in early developmental stages - an incomplete product with many irksome and quirky features. While they shared visions of carts rolling easily up and down hallways with a high degree of reliability and user friendliness, the current reality was quite different. I encountered considerable disillusionment but little pessimism. Pioneers in the use of this strategy, most of the schools were committed to making whatever local adjustments were necessary to reach their original vision.

At the same time, many of the schools were disappointed in vendors who had promised much but often delivered a product that was flawed in some respects. Creating reliable wireless network zones that could be counted to serve all classrooms with consistently strong performance had proven to be a bigger challenge than many had anticipated. And then, too, changes in system software would sometimes force a revision of a working solution.

The Team Schedule

Optimal use at the elementary level is likely to occur when pairs or trios of teachers in neighboring classrooms sign up for a laptop cart for several days or a week at a time. They then handle the hour-by-hour movement of the cart from room to room in a somewhat informal manner, shifting resources back and forth across a hallway as needed.

They have enough equipment at hand to accomplish something deep and worthwhile, but they also have neighbors who will take the laptops when they are no longer needed. The sharing takes the pressure off any one teacher to use the equipment all day long or feel guilty about glowing screensavers.

Scheduling and sharing within a single department works in much the same way at the middle school and high school levels.

Offering only brief time slots for individuals tends to encourage trivial pursuit and tangential activities.

Allowing individuals long chunks of time without sharing can prove wasteful and divisive, as even the most zealous teacher is likely to put the laptops aside throughout the day while discussions and other activities occur. Resentments can flourish when other teachers see the equipment languishing or being hogged.

The Unit Mandate

In order to prevent domination of equipment by early adopters and enthusiasts, curriculum units are adopted school wide. These units call for judicious, blended use of new technologies.

Perhaps the fourth grade science curriculum calls for a two week unit on acid rain, a two week unit on weather and a two week unit on simple machines. Unit plans created during the summer by teams of teachers are adopted as mandatory curriculum experiences. Blending the best of new technologies with the best of classical learning strategies, all fourth grade teachers know that they will need to schedule the laptop cart during these three units. The only question becomes which weeks will work best for each teacher and each class.

Experimenting and Adapting

A spirit of enthusiastic innovation persisted across the schools I visited. Even though mobile computing had proven much more challenging than these schools had expected, they remained eager and committed to the principles that had won them over to the strategy in the first place.

One of the goals of this book is to protect newcomers from the surprises and disillusionment sometimes experienced by early adopters. By sharing stories of issues and problems, the next wave of schools should be able to avoid weak batteries, heavy laptop carts and ponderous network software.

Chapter 6 - Why Not Share?

In some schools that have tried laptop programs requiring all students to buy laptops, the search is on for the next best thing.

What's next?

Some schools are looking at their belief statements about cooperative learning and they are challenging the value and wisdom of buying a laptop for each student.

Some of these early laptop schools may move forward to the strategic sharing of resources.

If laptop computers might be used less than 25% of the time, why not move them about? Why not buy one for every 4-5 students and then move them about, encouraging sharing by pairs and trios? Why buy a computer for each student and let it sit idle most of the time? Can we schedule strategically to optimize use, reduce cost, minimize waste and reserve funds for program and professional development?

Some independent schools are even using as a marketing advantage the fact that they supply (and share) computers instead of requiring parents to buy them.

Sharing and collegiality may be the wave of the future.

Many business and educational studies stress the value of team problem-solving skills. (National Alliance of Business. June, 2000) We look for teams of students to work together rather than operating in a solitary fashion.

The **Engaged Learning** model advocated by Barbara Means and others suggests four main elements, one of which is teaming:

Students will be . . .

- Responsible for their learning
- Energized by their study
- **Collaborative**
- Strategic

Why Not Share?

Strangely, when each member of a team possesses a laptop, the challenge of collaborating can be hindered rather than enhanced. Separate ownership can work against dialogue and synchronous exploration.

During a workshop with an independent school looking at the impact of different intensity levels of equipment, many teams reported that interpretive tasks often moved more smoothly and effectively when team members shared screens and laptops. (See workshop materials at http://questioning.org/laptop.html)

One advantage is economic. We can double or triple the impact of our equipment budget by reaching more students with fewer computers. We can also reserve funds normally devoted to purchasing equipment to generally underfunded elements such as professional development, program development, and technical support.

Another major advantage is standardization and maintenance of equipment to work effectively on the school network. When equipment belongs to the school, various kinds of games, instant messaging and other distractions can be pretty much eliminated. Reliability can be enhanced by providing an ongoing, systematic maintenance program.

If we plan strategically, moving laptop carts to where they can do the most good, we can provide a mixture of experiences. Sometimes we can have enough laptops on hand so that each student may work on an essay or poem on a single, personal laptop. At other times, students may team to consider an important, problem, challenge, or issue.

With proper planning, we can overcome the typical short battery life of laptops by providing extra batteries and charging facilities as part of the laptop carts, a real challenge to overcome when each student has their own laptop. Wireless too often becomes wired.

Sharing equipment is not new in schools. We have shared microscopes and balances in science labs. We have even taken turns using saws and drills and parallel bars.

There may be plenty of times when it is desirable for each student to work on a personal laptop, but that opportunity can be made available when and if it becomes desirable.

When we move computers about, we can enjoy the benefits of just in time equipment. It is an approach that conserves resources rather than squandering them. It is judicious and prudent.

When we make families (or taxpayers) buy computers for every child just in case students might need them from time to time, we run the risk of over-equipping our classrooms.

There are at least two troubling consequences. Teachers might feel pressured to make frequent use of these tools even when they should be closed and idle. On the other hand, the laptops may remain idle so much of the time that it will be hard to justify the large investment.

Why not gradually increase the number of laptops in a school as readiness and program needs develop? Maintain records of use. If high levels of use justify the purchase of additional units, so be it.

No laptop before its (it's) time!

References

National Alliance of Business. (June, 2000). Executive summary. **Building America's 21st Century Workforce**.

The 21st Century Workforce Commission believes that: The current and future health of America's 21st Century Economy depends directly on how broadly and deeply Americans reach a new level of literacy — "21st Century Literacy" – that includes strong academic skills, thinking, reasoning, teamwork skills, and proficiency in using technology. (Online)

When doing research, students should be more than hunters and gatherers. They should be capable of developing their own ideas.

Original thought is not a luxury or something we might expect only from the gifted. We need all students to learn to think for themselves and make up their own minds about the most important questions and decisions of their times and lives.

Democratic societies require citizens capable of challenging conventional wisdom, the propaganda of zealots and demagogues, as well as the platitudes and bland assurances of those in office who would like us to suspend critical judgment when systems break down.

New technologies can foster a cut-and-paste mentality and a new plagiarism. Students can be swept along browsing, grazing and collecting other people's ideas without taking the time to challenge those ideas or build their own.

Without critical and creative thinking capabilities, there is some danger that students will line up facing into the wind like sea gulls clustered upon an ocean walk.

Secondhand thinking is dangerous.

History warns us that masses are easily mobilized by demagogues who promise much and appeal to fears and anxieties. One antidote is a school program intent on raising students immune to such appeals.

At the heart of such a program is the frequent introduction of students to problem-based learning and decision-making - issues and questions that require fresh thinking during and after the research phase. Information gathering remains important, but it is something like the shopping for ingredients prior to the cooking of a good meal.

The student gathers the best herbs, onions, and tomatoes but doesn't stop with full shopping bags. The proof is in the cooking - the combining of ingredients into a stew that is unique.

The question is the answer.

Most new state curriculum standards stress four verbs . . .

- interpret
- infer
- analyze
- synthesize

The first three develop the understanding required for the fourth - synthesis being the invention skill.

Given a set of numbers, a collection of diary entries, a poem, an essay, or a photograph, the student is expected to read between the lines, crunch the numbers and find meaning.

Once they have interpreted and inferred, they are then expected to come up with new meanings - synthesis.

Australian children digging to China.

Inference is much like digging below the surface to find meaning.

Inference is detective work - seeking clues, resolving mysteries, figuring out whodunit.

The Internet is a great source of certain kinds of data and primary sources - images, diaries, and text files of various kinds. Often, meanings are implicit rather than explicit. The student must develop insight rather than uncover it.

But teaching this skill requires more than repeated practice. It is not enough to send students to an online museum and ask them to come up with new titles for paintings.

Teaching such skills requires the introduction of the thinking strategies that combine to enable students to extract meanings and to construct meanings when the raw ingredients are present but the actual meaning is not self-evident, followed by modeling and guided practice.

It turns out that success requires sustained focus upon the development of powerful questioning skills. Chapters Seven and Eight of this book propose an approach to support such learning.

The Illusion of Invention - Template Thinking

Just as painting by numbers is a far cry from real art, cutting and pasting the ideas of others is a long shout from real thinking. Collecting and harvesting is so easy for students these days that they may find themselves with baskets and disks full of ideas plucked from the pens and word processors of others, many of which they might not even understand.

Combining these mind bytes and idea fragments into a collage of thoughts (even when properly cited) is a poor excuse for thought. All too often, this combining is too glib, too facile and too detached from deep inquiry.

The Importance of Invention

Invention makes it possible to alter, modify, and shift things around us to make them better. But we should expect more than a shuffled deck of warmed-up old ideas and practices. We should see substantial value added.

Without invention, a society or a group rests on its laurels until it stagnates. Old truths stifle new ones. Instead of moving forward, revising and improving the way things work, we plateau, settle and coast along with little real commitment to change. The promises of earlier times rust and tarnish.

If a school, a corporation, a hospital, or a nation does not place a premium upon invention, then the prospects for a healthier and more productive future are severely limited.

At the same time, invention must be sufficiently grounded to produce results. It is not enough to be new simply to be new. We are not talking about pipe dreams and silly gadgets. We seek the kind of inventive thinking that will make a substantial, lasting contribution.

The next two chapters provide specific strategies to develop the kinds of thinking skills required to foster such inventiveness.

The building process begins with elements that may be skillfully combined into a new and different idea.

How do we teach young people to build their own good new ideas instead of simply cutting and pasting the ideas of others?

How do we show them the importance of inventing and of building on the best thinking of others while introducing new insights and possibilities?

How do encourage them to have faith in their creativity, their imagination, their freshness and originality?

At the age of four, children make up songs complete with lyrics and melodies. They rarely stop to doubt their talents.

At the age of four, girls and boys create towers of blocks that reach nearly to the sky. They even enjoy the tumbling down part of building. If their towers fall down, they scream with delight and start over again.

Same with sand castles.

Unconstrained by thoughts of performance, children can even improve upon the sky with wildly imaginative finger paintings.

The sky is their limit? Hardly.

They ask unanswerable questions like "How big is the sky?" and "Why do we have to die?"

If imagination and creativity are hard-wired into children from birth, where do they go? Can they be reawakened?

Chapter Sixteen explores the challenge of finding truth - the "difficult truth" as Michael Leunig refers to it. This chapter and the next are intended as companion pieces. The focus here is upon the building of **good** new ideas - the skills needed and the teacher strategies that may encourage the development of such skills.

We may need scaffolding to support the building process. It depends upon the question, the issue, or the challenge at hand.

Creating Good New Ideas 1

What is essential?

Most states in Australia and the U.S.A. are making bold statements about the kinds of thinking students must be capable of applying to the important questions, issues, and decisions of life.

1. Futures
2. Identity
3. Interdependence
4. Thinking
5. Communication

The above categories are drawn from the Essential Learnings statement in South Australia (SACSA). "Futures" is defined as follows:

"Students will develop capabilities to critically reflect on and shape the present and future through powerful uses of literacy, numeracy and information and communication technologies." http://www.sacsa.sa.edu.au/index_fsrc.asp?t=CB

Stating ambitious goals is one thing. The actual attainment of lofty goals is quite another matter. What does it take to push beyond conventional wisdom and create something new?

Seven Ingredient Hot Sauce

It takes (at least) seven ingredients to inspire Cajun thinking:

1. Guidance
2. Tools
3. Skills
4. Strategies
5. Faith
6. Playfulness
7. Resources

1. Guidance

By the middle years of schooling, most students will require the guidance of a mentor to show them just how to construct new ideas. This mentor (or teacher) will know how to suggest without telling, to model without overpowering, to encourage rather than direct. Because there are many paths to understanding, the teacher must not grab the student's mouse, figuratively or literally. Because discovery is essential to this process, teaching must subside, silence must prevail and questioning must be the source of direction and inspiration.

For teachers to be good at guidance, they must first wrestle with the challenge of creating new ideas in their own lives until they feel accomplished, experienced, relaxed and calm enough to support others.

One way to initiate a class to the construction of new ideas is to "walk" through one discovery experience as a group with the teacher leading the field trip, highlighting and modeling the process so that students have a chance to see firsthand how ideas are built.

This guidance role has implications for professional development. If we expect teachers to be guides in the development of ideas, we must provide ongoing, rich experiences for them as adults to do such thinking and exploration with mentors and teachers from the adult world.

Perhaps we team science teachers with a biologist, English teachers with a journalist, social studies teachers with a mayor and art teachers with a painter?

2. Tools

Each student needs a carpenter's belt to hold the mental equivalent of saw, hammer, level, wrench, pliers, and drills. While we might equip students with other technologies such as books, probes, microscopes, pencils, and laptops, the most powerful tools of all are mindware - human questions and problem-solving techniques that support innovative thinking.

Once internalized, these systems of thinking are embedded within student minds as frameworks for approaching mental challenges. They may operate for student brains somewhat as software can work for computers, except that they are often more dynamic and organic than most software programs. They interact, mutate, weave together, and shift with experience.

Once equipped with great mindware, when it is time for a mental saw, the student knowingly grabs a saw - a crosscut saw if a particular kind of conceptual cutting is required.

Technology cheerleaders can be a bit quick to push the merits of digital tools without linking them to appropriate mental tools. Without strong mindware, software and computers are unlikely to produce impressive new ideas, better student writing or performance.

This point is especially well made by John Seely Brown and Paul

Creating Good New Ideas 1

Duguid in their powerful book, **The Social Life of Information**.

   Hammering Information

   *Caught in the headlights of infologic, it occasionally feels as
   though we have met the man with the proverbial hammer to
   whom everything looks like a nail. If you have a problem,
   define it in terms of information and you have the answer.*

   (Brown and Duguid) Page 19

Mindware Example One - SCAMPER (Eberle, 1997).

Each letter of SCAMPER stands for an approach to invention.
When a student looks for a solution to a social or environmental
problem such as acid rain, she or he can take all the elements of past
efforts, mix them up with plenty of newer options and then apply each
of the techniques below to create a new plan of action.

   S=Substitute
   C=Combine
   A=Adapt
   M=Magnify, Minify
   P=Put to other use
   E=Eliminate
   R=Reverse

Mindware Example Two - The Questioning Toolkit

   First published in **FNO** (http://fno.org/nov97/toolkit.html) - a
collection of different kinds of questions that might be combined to
create answers to complicated and demanding questions.

Essential Questions	Subsidiary Questions
Hypothetical Questions	Telling Questions
Planning Questions	Organizing Questions
Probing Questions	Sorting & Sifting Questions
Clarification Questions	Strategic Questions

Creating Good New Ideas 1

Elaborating Questions
Inventive Questions
Irrelevant Questions
Irreverent Questions

Unanswerable Questions
Provocative Questions
Divergent Questions

© 1997, Jamie McKenzie

The student learns to switch from one type of question to another depending upon the type of mental operation required to build the next level of understanding.

Mindware Example Three - The Scientific Method

(or one of many creative problem-solving models available online or in book form.

Stages of Problem-Solving

Define Problem
Gather Data & Explore Possibilities
Invent Options
Evaluate Options
Create a Plan
Act

Examples:

• The Team Engineering Collaboratory http://www.vta.spcomm.uiuc.edu/PSG/psgl4-ov.html
• Creative Problem Solving Books from the Center for Creative Learning http://www.creativelearning.com/creativeproblem.htm
• Creative Problem Solving Bibliography from the Center for Creative Learning http://www.creativelearning.com/bibliography.htm
• Increasing Your Expertise as a Problem Solver: Some Roles of Computers. Moursund, D.G. (1996). Includes an excellent list of resources. http://www.uoregon.edu/~moursund/PSBook1996/

Mindware Example Four - Whack

The innovative strategies suggested in the books of Roger von Oech, such as **A Whack on the Side of the Head**. New York, Warner

Creating Good New Ideas 1

Books,1998. von Oech provides activities to unlock the mental locks that block innovative thinking. He offers strategies to inspire and support dynamic thinking.

Mindware Example Five - Thinking Hats

In his **Six Thinking Hats** book, Edward deBono explains how each of six differently colored hats can stand for a kind of thinking. The blue hat, for example, stands for thinking about which hat (or type of thinking) should prevail at various times. The green hat stands for creative thought. The yellow hat stands for positive thinking - looking at the advantages. White hat thinking is careful and objective collection of data. The red hat requires attention to emotional issues. A sixth hat (I prefer the color purple for this one) stands for skeptical, doubtful, and critical thinking. All six types are important, but it is the conscious orchestration of these six types of thinking that makes productive and imaginative thinking possible. The hat metaphor puts such orchestration within the reach of elementary school children.

Mindware Example Six - ThinkerToys

Michalko's book, **ThinkerToys**, is a well organized collection of dozens of problem solving strategies and tools.

Linear Thinkertoys
False Faces (Reversal)   Slice and Dice (attribute listing)
Cherry Split (fractionization)   Think Bubbles (mind mapping)
SCAMPER (questions)   Tug-of-War (force-field-analysis)
Idea Grid (FCB grid)   Phoenix (questions)
Future Fruit (future scenarios)   Brutethink (random stimulation
Ideatoons (pattern language)   Clever Trevor (talk to a stranger)
Idea Box (morphological analysis)
The Toothache Tree (diagramming)
Hall of Fame (forced connection)
Circle of Opportunity (forced connection)
The Great Transpacific Airline and Storm Door Company (matrix)

Intuitive Thinkertoys
Chilling Out (relaxation)   Blue Roses (intuition)
The Three Bs (incubation)
Rattlesnakes and Roses (analogies)

Creating Good New Ideas 1

Dreamscape (dreams)   Da Vinci's Technique (drawing)
Not Kansas (imagery)   The Shadow (psychosynthesis)
Stone Soup (fantasy questions)
Color Bath (creative visualization)
The Book of the Dead (hieroglyphics)
Dali's Technique (hypnogogic imagery)

Group Thinkertoys
Brainstorming Rice Storm (TKJ)

Endtoys
Worrywillie's Guide to Prioritizing
Murder Board
Backbone

Mindware Example Seven - Mental Mapping and Cluster Diagramming

Students can learn to map out complex issues and questions. They grow accustomed to visual representations of ideas and the links between them. Comparing ship captains, they begin by identifying criteria to guide choice and then they develop a clear picture of the prime questions needing exploration.

Such mapping can be done on paper or a computer screen, or both. The most important aspect of this process is not the implement or physical tool (paper, pencil, or laptop). It is the visual and conceptual thinking that matters, and that kind of thinking must be grown

over time through modeling and instruction by a savvy mentor or teacher.

Navigation Skill?

- Did he know how to use all the best instruments of his time?
- Did he keep a careful log?
- Did he usually know where they were?
- Did he ever get lost?
- Did he seem to know what he was doing?
- Did his ships have to wander around very much?
- Did he stay clear of known hazards?
- Did he know how to make the best of prevailing winds?
- Did he know how to maneuver during a sea battle?
- Did he have mates that could help him when he needed it?
- Did he know when to ask for directions?

The tools role has implications for professional development. If we expect teachers to show students how to wield such powerful thinking models and tools, we should be proving those teachers with 30-45 hours of coursework in several of these models until they become comfortable and proficient.

3. Skills

Closely related to the thinking tools mentioned above are the skills required to apply those models and tools to actual situations. It is not enough to acquire the models in the abstract, apart from actual problem-solving.

Ownership of a powerful drill does not automatically confer upon the owner effective drilling. One must know something about selecting the right length, diameter and composition of drill bits to match the task. Once selected, it takes some skill to load them and apply pressure in a firm, effective manner.

Without these skills, the apprentice uses a bit meant for wood to drill through concrete and finds the drilling difficult and the bit dulled.

When it comes to ideas, the effective use of a cluster-diagramming software program like Inspiration™ requires dozens of seemingly minor skills that may have a significant influence upon the generation of ideas. Assigning colors to particular concepts and dragging them to their own section of the diagram can add a level of coherence and intelligibility to the exploration that exceeds one's normal expectations for the impact of coloring. Assigning symbols to match concepts, questions, or components can also elevate the impact of the diagram.

The effective use of graphic organizers and other thinking tools takes extensive practice so that the component skills become pretty much automatic. Early efforts can be clumsy and stiff, hindering the thought process rather than enhancing it.

Balance of Skills: To optimize results, students must develop a high level of skill across all of the major categories of tools (mindware, software, and hand tools) listed earlier. Training in spreadsheeting does little to enhance student performance without a balanced commitment to showing students how to interpret numbers and communicate visually about numerical relationships. Training in powerpointing does little to enhance student performance without a balanced commitment to showing students how to build ideas and communicate visually with attention to aesthetics. It leads instead to powerpointlessness. (see "Scoring Powerpoints, the September, 2000 issue of **FNO**.)

The Visual Display of Quantitative Information by Edward R. Tufte ISBN: 096139210X Graphics Press, 1992

4. Strategies

Creating good new ideas requires the strategic use of the tools and the skills mentioned above. There are no recipes, templates, or automated procedures that are likely to foster or nurture imaginative production. Strategic use entails reflection and choice. There is often some trial and error, some play, and some experimentation.

- Which tools will work best here?
- Which skills do I need to apply?
- In what order should I proceed?
- Have I asked the right questions?
- Do I know what I need to know?
- How can I best combine tools and techniques for this particular challenge?
- What is missing?
- What is needed?
- How can I do this?
- How would my models/heroes do this? my opponents? my competition?
- Which approaches that have worked for me in the past might work here?

- Have I anticipated what might go wrong?
- Have I planned for surprise?
- Do I have a flexible plan?
- What else do I need to consider?
- Where can I seek inspiration?

5. Faith

A surprisingly strong determinant of creative production is the faith that one can add value by applying one's best thinking to a challenge. Sadly, a relatively small percentage of young people seem to emerge from school with this faith. They pass through too many discouraging experiences that communicate the opposite message.

While there are some models of gifted education like the Renzulli approach that define giftedness broadly and set high expectations for all young people, many students seem to pass through school thinking that the writing of songs, poems, essays, and music is the special province of a very small percentage of the population. The same with ideas. The message seems to be that a tiny fraction of the population creates new ideas while the rest are condemned to memorizing the ideas and insights of others.

Catford and Ray identify faith as one of the four most important tools of the hero in their book, **The Path of the Everyday Hero**.

6. Playfulness

In addition to the confidence mentioned in the previous section, inventors need to approach challenges with humor, laughter, flexibility, and fun. They must be able to turn down the voice of the critic within and allow their clowns a chance to play. Great new ideas often seem silly when they first surface. We are pushing out the walls of our previous experiences and dancing with fanciful new possibilities. We must do so with open minds and a spirit of whimsy.

Research on the creative process by Torrance and others has shown that measured creativity often drops beginning at the fourth grade for two main reasons:

1 Right answer teaching
2 Peer pressure

Torrance found that students reduced their divergent responses and productivity as they tried to narrow their focus on figuring out

64

what their teachers expected. He also hypothesized that young people were reluctant to offer ideas publicly, suggestions, and insights that might be viewed as weird or dumb by their peers.

It is unlikely that the possession of a laptop or a hammer will do much to break these patterns by themselves. If we hope to see more playful students with more divergent responses and creative production, then we must create schools and classrooms that honor playfulness and imagination.

Reaching such an ambitious goal would require fundamental changes in the culture of most schools, beginning with the way teachers work together. The schedule of most schools is so tight that there is little room for imaginative play or invention. Michael Fullan questions how much change we can expect when teachers are preoccupied with what he calls "the daily press."

The daily press (the need to take care of moment to moment classroom pressures) is a mammoth obstacle to be overcome if an innovation is ever going to take root. Teachers are often constrained from thinking about new ways of organizing learning in their classrooms by the need to handle day to day issues, surprises, crises, and challenges (Fullan's summary of research by both Huberman and Crandall, pp. 33-34). This daily press creates the following impacts:

- enforces a short term perspective and an emphasis upon coping
- isolates teachers from dialogue with colleagues
- exhausts them - leaving little left for special efforts and sprints
- limits opportunities for reflection
- makes them dependent upon what they already know and prone to following routines

Fullan, Michael G. (1991) **The New Meaning of Educational Change**. New York: Teachers College Press.

7. Resources

Ideas thrive, flourish and multiply in richly fertilized and well cultivated environments. Fallow fields produce few sun flowers.

While we often criticize the information available on the Net for its lack of reliability and quality, the new information landscape can its lack of reliability and quality, the new information landscape can provide a stimulating stew of provocative sources capable of stretching our minds and inspiring new ideas. (see "Learning Digitally" in the November, 1998 issue of **FNO**)

Fresh perspectives and wide-open windows to the seething new electronic bazaar of ideas could promote fresh thinking unlike anything we have witnessed in the past.

Innovation and originality thrive on the free flow of ideas and experiences. At the same time, much of the new information and the electronic media suffer from clip art banality and a template sense of style. Just as much of the world has been GAPped by the mass marketers, the news wires and McDisneySoft empire undermine creativity with their unrelenting drive toward standardization and "ready made" ideas.

The Web allows mavericks, clowns, heretics, poets, and fools to publish their work without bowing before the editors, sages, and elders who have so long dominated the flow of ideas. This is a decidedly mixed blessing, of course, because it means the startling insight and bold dash of color may be submerged in a flood of mediocre and disappointing offerings.

As with the other potentials we can ascribe to digital learning, the prospects for a surge of creativity and originality will require some promotion and catering. None of this will happen automatically. While some idea generation will thrive spontaneously like a virulent virus, much of the good will be offset by countervailing viruses spawned by mass marketing and mass media. Clip art, templates, and user-friendly short cuts will undermine some of the best prospects.

There is much that schools can do to enrich the learning environment, investing, for example, in the creation of new vertical files - digital collections stored on district file servers to support creative thought and inquiry. (see "The New Vertical File: Delivering Great Images and Data to the Desktop" in the October, 2000 issue of **FNO**).

The finished product - a new idea, fresh, compelling and possibly even inspiring.

If we build carefully, we can pull down the scaffolding that guided our construction to enjoy the finished product.

The South Australian government, like many other states and provinces, provides a list of essential learning goals that is quite a challenge for schools.

SACSA lists five major categories of performance:

- Futures
- Identity
- Interdependence
- Thinking
- Communication

"Learners develop capabilities to critically reflect on and shape the present and future through powerful uses of literacy, numeracy and information and communication technologies."

How can we best teach our students to think in this manner and create good new ideas?

Building new ideas is somewhat like constructing a grape arbor in an unusual location requiring custom installation, inventiveness, and an appreciation of the impact of a new structure on the larger context.

Big piles of facts and information do not contribute much by themselves. The challenge is building something new from the raw ingredients.

For the past few years, we've been hearing a bit too much about information and not enough about meaning. (see review of **The Social Life of Information** - http://fno.org/may2000/review.html)

Brown and Duguid are insiders. They spend their lives where information technologies do their best and their worst. Fully acquainted with the hype and promises of information cheerleaders - a group they call infoenthusiasts - Brown and Duguid warn that ". . . it can be easy for a logic of information to push aside the more practical logic of humanity." (p. 18)

They fear that an obsession with information can lead to a kind of tunnel vision with planners ignoring much of what lies within the periphery.

There needs to be more emphasis in schools and elsewhere about converting information into something that makes a difference in life.

Students should learn how to translate data and information into insight using analysis, interpretation and inference skills. But we should not stop with understanding and insight. We should also show students how to turn that understanding and insight into information products of various kinds: inventions, decisions, solutions and proposals meant to improve society or enterprise in some fashion.

The student converts information into something practical, useful, and novel.

How can we best protect beaches and wild species from pollution, oil spills, and other threats?

Unless we shift school research away from topical research, there is little chance students will learn to form new ideas, create solutions to pressing problems or make well-informed, thoughtful decisions.

Finding out about a foreign country, a province, a state, a general, a battle, a chemical substance, or a social issue does little to sharpen thinking skills. Topical research requires little more than collection of information.

Creating Good New Ideas 2

When we ask students how we can best protect or restore an endangered species or habitat, or when we ask them to suggest ways to breathe life into an industry that is struggling (such as some online businesses, some fisheries, and some brick and mortar retail outlets), then we challenge them appropriately.

They should not be able to find an answer. Cut-and-paste thinking should fail them. They cannot collect conventional wisdom. They must come up with fresh answers.

For more on the process of building studies around essential questions, read "The Question is the Answer: Research Programs for An Age of Information" - http://questioning.org/Q6/question.html.

1. Considering Conventional Wisdom

There are many ways to keep grapes off the ground. Some of these work better for vineyards raising wine than homes with no intention of crushing grapes into wine or juice. The above shows New Zealand grapes awaiting harvest for wine.

Whether building a grape arbor, crafting the solution to an environmental problem, or fabricating a new Web site, an invention team should definitely consult conventional plans, conventional wisdom, and leading authorities to see which strategies have been tried in the past, which elements have proven successful and which aspects might require a fresh approach.

This opening stage of research is all too often the final step for teams without faith in their own creativity, originality and abilities, but progress requires something more demanding and more exciting.

We are looking for value added, the next generation of mousetrap or mouse or track pad. We are committed to the notion that human ingenuity can usually find ways to enhance and improve what has gone before.

2. Adapting Plans to Local Conditions

Before construction of the grape arbor: a row of vines that had been held up by posts and wires until a storm blew the rotten posts down.

While one can find plans, schemes, strategies, and packaged programs to match almost any challenge, we have plenty of evidence that change takes root in the healthiest ways when the innovation is adapted to match local conditions.

Saving the salmon in a river in Oregon will be different from saving salmon in Washington, British Columbia, Maine, Tasmania, or Scotland. The rivers are different, the salmon are different and the elements of the plan must be matched to special combinations of factors.

Introducing wireless laptop carts to middle schools in three different towns in Oregon with three different groups of students and teachers will work best if each school creates its own plans matched to special conditions. The same would hold true for laptop carts in elementary or high school classrooms in Oregon or middle school classrooms in Tasmania, Dunedin, Blackpool, Dallas, Sydney, or Victoria.

In the case of the grape arbor project outlined in this article, the grape vines were pre-existing and quite old. They were situated along a curved terrace that would make a single straight arbor an impossibility. The setting included several other levels of terraces, each with different kinds of plants, some of which might be hidden by a structure too tall.

<-- Terraces before Arbor Project

The arbor had to look good from the bottom corner of the property looking up through the terraces, but it also had to look good from the house looming above. The house looked down upon the terraces from three levels - the top floor, the middle floor and the ground level at the back of the house that held several other gardens.

It would have been fairly easy to buy several prefabricated units. It would have been easier to copy the linear post and wire design of the previous owner. But it was time for a change - an arbor that was customized to match the curves and special qualities of the terraces . . . an arbor that would compliment the thriving lavender, rosemary and Russian sage plants on other levels - an arbor that would echo the theme set by the four half wine casks that offered bright splashes of color throughout the hot summer months.

I found a photograph of a single grape trellis with a ladder top and decided to build three sections following the basic plan with modifications to fit the length, height, and shape of the terraces.

3. Collecting the Elements

Lumber laid out on the driveway ready to be cut into the right lengths to match the plans. Not shown, the cement, the galvanized screws, and the many tools required to finish the job.

The lumber and the cement required for this job were too large and heavy a pile for my Volkswagon Jetta's small trunk, so I had the pile delivered by a very large truck that carried its own fork lift. The driver skillfully lifted my pile from the truck and left it close to the project site.

Technology at its best!

When preparing a good new idea, we must also go on shopping trips to find the main ingredients and key elements of our final proposal. We may find them in their raw, unfinished forms, but the building of ideas requires collecting, piling, sorting, and storing information in ways that it can be synthesized later to create the good new proposal.

Are there times we would welcome an information delivery or a fork lift? Yes, indeed. But much depends upon our ability to order just what we need. There's not much to be gained by someone dropping off a huge pile when we need highly select, very pertinent information.

4. Digging the Holes for a Foundation

We are often warned to build our ideas, our houses, and our relationships on strong foundations. The prior system of wires and posts fell down in a wind storm. To dig thirteen holes 2.5 feet into the ground and 2.5 feet in diameter, I rented a large augur.

When we search for new solutions to old problems, we should avoid using the rotten wood, shallow thinking, and flawed planning of earlier attempts. We arrange the right tools for the occasion and dig deeply and well, determined to build our new system on firm footings. We prepare the ground. We cultivate. We don't rush things, pounding our stakes into the earth as if mere force will suffice.

The trellis book I had consulted underlined the importance of digging holes and setting concrete. The author also recommended the use of power tools for such a task.

Such good advice. And yet here was a demanding set of new skills and a tool that required new learning for me to succeed.

It is the same way with research. Sometimes we cannot access the information we need or convert it into meaning until we have acquired new tools and new skills. Data may be readily available while meaning lies far below the surface.

In looking to restore species and habitats to healthy conditions, for example, researchers must do more than end the current threats. They must dig back into time to figure out what conditions existed before the degradation of environments began.

Effective solutions are almost always deeply rooted in past experience and tied firmly to evidence that supports and validates the strategies being proposed.

5. Resting

Sometimes we must rest from our labors, knowing that great new ideas are nurtured by our dream space. We also rest to regain strength for our next challenges.

It was a very big and very heavy machine. Each time the augur drills down into the soil 6-8 inches, the operator must lift it back up to the surface so the soil gathered in the spiraling coil will spin to the sides.

Heavy work. But so is thinking. Heavy mental lifting.

Sometimes we put ourselves in time-pressured situations where there is no room left for rest or recovery. This is folly. Invention thrives on reverie and calm reflection. If we don't set aside time for deep thought, we are much less likely to discover great new possibilities.

6. Assembling and Cementing

The idea was to take the 4"x4" posts and tie them together across their tops with 2"x4"s. These could then be planted in the holes and cemented into place before adding the ladder like top structures.

Chunking is a favorite problem-solving strategy of many people - breaking up a task or problem into smaller sub tasks and then putting the chunks together later.

Sometimes people make the mistake of putting all the pieces of a new idea together before lifting it off the ground and testing it under realistic conditions. The new idea may be great in theory but sad in practice . . . weighty and impressive but impossible to sustain.

In this case, it was hard enough to carry six large posts and their framing down to the holes in the terrace. It took two people to carry the structure down and there were several times when it threatened to fall over and break.

Could the same happen with new ideas?

Thinkers and inventors learn to stage the development of projects and ideas, being careful not to rush the process or overload the system.

7. Foundation Work

The cement was intended to protect the wood from rotting by limiting contact with damp soil. It was also intended to hold the posts in place against the weight of grapes and the blasts of winter storms.

When building new ideas, we must be careful not to set any of our posts in concrete too early in the process. While this strategy may work fine for grape arbors, ideas should properly remain quite fluid until the very final stages of refinement and development. Premature fixing of ideas can block new possibilities and combinations from consideration.

Even when building grape arbors, it pays to leave the wooden structure sitting uncemented in the holes propped up to allow the planners a chance to evaluate the look and the feel of the new structure in its setting. How different it looks in the actual ground.

Once the concrete is set, it will take a chain saw to change the height or position of the structure.

Some people approach invention and the creation of new ideas with so much bias, predisposition and closed-mindedness that they are doomed to failure. Strong foundations can support the lofty spires of a cathedral, but they can also a prison make.

When we gather information to support our thinking, we don't mean selective gathering. The goal is to gather all the information we need to build a clear understanding.

Sometimes the best inventions are carefully grounded in the consideration of everything that might possibly go wrong. Pie-in-the-sky thinking is fine early in the process, but implementation plans require attention to obstacles, issues, threats, and surprises.

8. Assemby

Once the large posts for the first section were cemented firmly in place, leveled and carefully squared, it was time to construct the ladder frames that would rest on top of the posts and hold the grape leaves and vines.

A really complex idea or proposal may contain hundreds of details that must be organized and structured in ways that contribute to understanding. The glory may lie in the bold strokes because they seem dramatic, but real triumphs sometimes rely upon carefully orchestrated campaigns involving many small steps.

The Tipping Point: How Little Things Can Make a Big Difference (Malcolm Gladwell, 2001) tells of many campaigns that relied upon the little things to make a huge difference. He gives as an example the ending of turnstile jumping in New York subways as the intervention that led to a vast decrease in the violent crime rate.

If we ask students to solve the problems of the Snake River, it may be tempting for them to grab the simple, sweeping solutions. Take down all the dams! Leave all the dams up! End all fishing!

But the most powerful solutions may involve much more subtle campaigns that require adjustments of fishing seasons, permitted catches and job retraining incentives, campaigns that involve modification of certain critically important dams instead of all dams, and campaigns that invest in alternative sources of water and energy to reduce reliance upon Snake River sources.

How do we teach young people the value of patient work with complexity? We give them practice with real problems so they can see firsthand the benefits of complex programs of action. We ask them to study the lives and actions of those who have modeled this kind of problem solving.

Brick by brick, we teach them to build a wall. We also teach them, brick by brick, how to dismantle a wall.

9. Learning New Skills

There are often some thinking and learning skills we may not possess when we begin our project. We wade in full of confidence, only to hit a dead end, break a tool or find ourselves "at wits' end." What choice do we have? We can quit then and there, or we can figure out a new way, learn a new skill, and get on with the project.

Sometimes we don't know what we don't know. At other times we may not know what skills are missing from our repertoire but required for success. We may not discover what is missing until we fail, fall flat on our faces, and hit bottom.

Old models of research placed a premium on finding and collecting. Invention requires much more. We must still be skilled at learning what we need to know, but we must also be skilled at knowing how to learn new ways of learning and thinking as we proceed. We must be good at learning new skills on the run, on the job, in the trenches, or wherever!

10. Synthesis

The first set of posts stood in the cement awaiting the arrival of the ladder section that would sit on top.

Synthesis is the rearrangement of parts and elements into a new combination or version. In the case of the arbor, we began with many pieces of wood and started assembling them into sections, each of which would eventually find a place in the overall structure.

The building of a new idea, a solution, or a proposal can proceed in a similar fashion. A plan to deal with the issues of the Snake River might be made up of many different sections, each of which might have impact on the others. There might be a section for modifications to dams and another one for irrigation issues and strategies. A dozen or more sections and plans will eventually stand synthesized into one comprehensive action plan.

11. Considering Context

How well did the new structure fit into the pre-existing context? How should the design of the second and third sections be influenced by the look of the first section. Even after plans have been drawn, there is still room for change as the real structure emerges.

One of the biggest failures of builders, architects, planners, thinkers, and innovators is the neglect of context. Installation of new equipment, new buildings, new programs and new ideas can falter and fall short of expectations when abstractions are converted too rapidly into action plans and actual implementations that conflict with local conditions in ways not foreseen by the authors and thinkers.

Ivory towers, think tanks, and incubators are all vulnerable to this trap. Removed from real conditions, the thinker may spin out the threads of dreams in ways that seem plausible and attractive in the conception phase but shrivel in the light of day.

Frequent, periodic reality testing is the antidote. New ideas must be subjected to pilot testing, peer review, and examination in context.

Was the new arbor too tall? Should the next sections be just as tall? Would they block the view? Should they be shorter? Should the ladder section be eliminated? And how would it look once the grape vines rose to the top?

Some of this testing can occur in real space. Some might require imagination and simulation.

12. More Assembly

The first section of ladder is ready. By accident, the new pieces are laid upside down on top of the posts below. A good accident? A lucky design change?

As we work on ideas and proposals, sometimes we discover new angles and new possibilities almost by accident, just as the discovery of penicillin is said to have followed the surprise discovery of spores that had grown overnight.

The inventor and thinker learns to welcome and encourage these kinds of surprises, maintaining an open mind and a playful spirit, relaxing enough to see how things might be different if we reversed some element, or replaced it with its opposite, or turned it upside down or inside out.

The key element in all of this inventing is the dynamic and organic nature of ideas. While we might strive to constrain, control, and shape ideas to match our wishes and preconceptions, this often proves dangerous and self-defeating as we cut ourselves off from a discovery process that might be magical and inspiring.

Logical analysis has its place, but it can be a blunt, counterproductive instrument incapable of influencing growth in a harmonious manner.

13. Combination

The second section of ladder is added to the top. Three parts have now been combined - the post structure and two ladder sections.

In building ideas, action plans and solutions, we may have a number of sections that require melding and orchestration. While it is tempting to paste these parts together in a simple cluster or sequence, the planner must be alert to interactions and complications.

How well do the elements fit together? Do the disparate parts of the idea or plan match up harmoniously, or do they need to be adjusted for a better fit?

The old approach to school research with its emphasis upon collection did not challenge students to think about melding. It was enough to list ideas, strategies and possibilities. The whole issue of orchestration, balance, and interaction was rarely addressed.

Preparation for problem-solving in real world contexts requires more than formulaic listing of actions. Strategic implementation requires adjustment and finesse.

14. Revision

When the two sections of ladder were laid atop the posts, a gap appeared on one side. A bend in one long side had created the problem. What to do?

Stuff happens despite the best intentions of planners and thinkers.

In the case of the arbor, the solution was simple - fill in the gap with a small piece of wood to hide the mistake.

In the case of real research, the solution may be much more elusive. Filling in missing data might be unethical and create false results. The gap may be an important discovery - an anomaly.

When conducting research or building new ideas, the thinker must be open to change and revision, sometimes even reversing direction entirely. As pieces of ideas and concepts are laid side by side, new understandings may emerge, and the thinker should entertain those new possibilities, adjusting her or his thinking in response to observation and learning.

15. Completion?

How do we know when we are done? In the case of the arbor, there was still more ground to cover, still more vines to support and plenty of lumber and concrete to build more structures. Knowing when to stop when building ideas is a bit more complicated.

The building of good new ideas can seem an endless process. If it were not for deadlines and pressures to take action, we might go on tweaking our plans for months and months. Ideas can ferment and change with reflection and experience, deepening and improving with time and consideration. Unfortunately, few of us have that kind of time or patience. We are apt to rush toward implementation rather than stepping back and letting time give us perspective.

"Fools rush in . . ."

One way to determine readiness is to establish some standards ahead of time.

This project will be ready for implementation once we have satisfied all of the following conditions:

- We have successfully field-tested the main strategies.
- We have consulted leading authorities in the field, and their suggestions have been included.
- We have considered and tapped all important sources of new ideas and unusual thinking.
- The project plan is comprehensive, clear, and coherent.
- Etc.

16. Extension?

Sometimes we make a plan and stick to it even when evidence accumulates that modifications are important. We stop learning and start implementing. We become so intent on drilling and filling holes or classrooms that we do not stop to notice danger signs or evidence that the plan is not working.

Ideas, plans, and proposals are meant to improve things. If those who suggest them do not take note of the true quality and worth of a project or an idea when it is set in motion, there is a serious risk that the project will fail.

If we teach students that all good ideas and projects are cooked up far from the field of action or the real context, we do them a real disservice. The best plans are modified as they are put into place.

17. Synthesis

What are our choices?

How can we make reasonable changes to improve things? Synthesis does not require drastic modification.

When we are building an arbor, we can change height and complexity, and color and location. Small changes can have gratifying effects.

When we are exploring ways to solve the problems of the Snake River, some people fall into simple camps of removing or leaving dams, while others realize that some dams are more of a problem than others, and still others find ways of partially removing dams or reducing their damage.

18. Two More Sections

The completion of each section serves as a warm up for the next sections.

Most of us can think of projects and problem-solving in terms of stages, but it is less comfortable applying this same mental framework to the development of a good new idea.

Some people build houses in stages, one wing at a time, over several generations. Could we do the same with ideas?

Some of us have been exploring how we might best launch engaged learning with new technologies, but our ideas on how best to achieve this goal keep shifting as the tools and the contexts evolve.

In March of 1998, I wrote "The Wired Classroom: Creating Technology-Enhanced Student-Centered Learning Environments" - http://fno.org/mar98/flotilla.html.

By January of 2001, that same idea had evolved to "The Unwired Classroom: Wireless Computers Come of Age" - http://fno.org/jan01/wireless.

When ideas must make their way into the world and be tested by reality, they are apt to pass through stages and generations of development.

If ideas are not adjusted from time to time, they creak with age and irrelevance like the outmoded science concepts and information sadly retained in many old textbooks and library books still in schools.

19. SCAMPER

One model for making change is a set of synthesis strategies called SCAMPER (Eberle, 1997), with each letter standing for a strategy.

S=substitute.
C=combine.
A=adapt.
M=modify, magnify, minify.
P= put to other uses.
E=eliminate.
R=reverse.

The thinker arranges, blends, combines, integrates, tests, and adjusts the thought fragments until new pictures emerge.

These strategies work well to inspire thought about modifications to the arbor. They also work well when adapting an idea, a proposal, or a project.

Creating Good New Ideas 2

20. Wondering

Sometimes we need to step back from a project to allow our sense of wonder room to play. If we are too fixed on performance and production, we may not discover much that is novel, imaginative, or vibrant.

Wondering is closely associated with reverie, dreaming and imagining. It is part spirit, part state of mind.

"I wonder what would happen if . . .?"

"Suppose I made this part darker and this part bright orange?"

"Wonderful!"

21. Looking Around

Sometimes it helps to wander further afield, step off the job site,and travel to some exotic location that might stimulate new thought and provide new perspectives.

We are all at risk when it comes to producing new ideas. To some extent our surroundings, our groups and our organizations apply pressure for us to conform to certain expectations and honor certain boundaries.

If we are building something like an arbor, perhaps we climb aboard a ferry and visit some nearby islands to see how they've been building arbors.

If we are building ideas or new strategies of some kind, we climb aboard an airplane to see what folks have done on the other side of the mountains or the continent or the ocean.

We may sometimes delude ourselves into thinking our way is best. We stay close to home and pat ourselves on the back.

"The American way is best." "The New York way is best." "The Australian way is best." "The Magic Marvel School way is best."

But then we cross a boundary, step across the street, pass through the Looking Glass and discover how little we knew after all.

In my own case, frequent visits to Australia and New Zealand have taught me much that was worth bringing back to Americans. "The Information Literate School Community" is just one dazzling example of many great ideas springing up in the land of OZ but well worth carrying home.

Without crossing the ocean, I have found the same phenomenon to be true when crossing state boundaries. What are folks doing and learning in Mississippi, Texas, Iowa, Nebraska and Michigan? What are teachers doing with rolling laptop carts in affluent suburbs? in big city districts? in independent schools?

We learn by looking past the obvious, figuring out what has worked and what hasn't worked so we can create our own plans without repeating the same mistakes.

Can we teach our students that the best ideas may not be found in their own backyards? Can we develop a respect for the inventiveness and ingenuity of other lands and cultures? Can we inspire students to cross boundaries to learn other languages and ways of thinking? Can we cure them of narrow mindedness and ethnocentric perspectives?

Sometimes we want them to think in an Eastern manner. Sometimes we hope they will think in a Western manner. At other times we hope they will think Southern or Northern or compassionately or skeptically.

We know that narrow-mindedness and prejudice are dance partners whose favorite steps cause much misery and pain. Can we teach them to build good new ideas that incorporate decency, consideration, balance, and concern?

Conclusion

The project evolved from one stage through two more, with each section a bit shorter and a bit lower than the first. After the arbor was in place, the grape vines climbed skyward and added their own charm to the project, some of which could be shaped by pruning and some of which was a bit random.

The intention of this chapter was to explore the development of an idea or a project through various phases of thought and inquiry. All too often, school research seems rushed and little concerned with the creation of useful new ideas. Ideas take time and care. If we rush toward decision, we may decide poorly.

If we expect to raise a generation of thoughtful decision-makers and problem-solvers, we should be taking a new look at the way we assign school research - not only the types of questions we assign but also the kind of research and invention process we encourage.

How can we best change school research so as to raise a generation of thoughtful decision-makers and problem-solvers?

Chapter 10 - The True Cost of Ownership

When most schools and districts install networks, they are unlikely to appreciate fully the wide range of costs accompanying the installation. Some of these will be dollar costs. Some will be organizational and psychic. When major costs and impacts arrive unannounced and unanticipated, they may cause embarrassment and damage. It makes good sense to look ahead without rose-tinted glasses.

One way to expand awareness is to adopt an approach from business called "The Total Cost of Ownership" (TCO) - a model that is well documented and translated into school terms by CoSN, an organization that supports schools in their efforts to bring networks into schools in ways that will actually enhance learning. (CoSN presents strong TCO support at http://www.cosn.net/.)

The main thrust of TCO is to identify all of the primary elements required to make any innovation succeed, no matter what the organization. Each of these elements, so the theory goes, may be essential, much like key stones in a curved archway. If we pull out one or two stones, we know the entire archway is likely to come tumbling down, or in the case of innovations, fail to lift off the ground at all. Applied to the launching of innovative programs, TCO suggests that failure to fund several key elements may fatally undermine the success of the venture.

Valuable as TCO may be as a tool to increase district awareness, this chapter suggests that TCO as it is usually applied to school technologies by CoSN and others does not adequately address other critically important costs that could have major implications for district and school success in broader terms.

For this reason, this chapter proposes a broader view - one that lists and describes all of those missing TCO elements without which the educational technology innovation may fail, sputter and even do damage to the educational programs of a district. It is meant to supplement and enhance the good work already done by CoSN on this issue.

The True Cost of Ownership

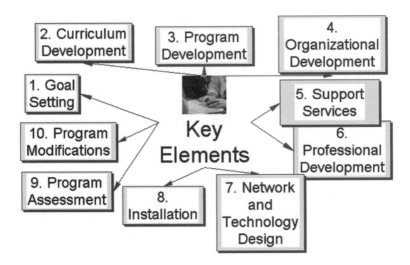

The Key Elements

In **Planning Good Change** (McKenzie, 2001) I outlined a model for technology planning that put student learning and curriculum at the front of the planning process (see figure above). In a chapter of that book, "First Things First" (also available online at http://fno.org/nov00/f1.html), I explained how school and district leaders may orchestrate those key elements in a cycle of invention, implementation, testing and adaptation.

We have learned in the past decade that the installation of equipment is a relatively straightforward task. The winning of enthusiastic support and the daily appropriate use of these new tools by classroom teachers is quite another matter. Few districts have managed to identify or fund the robust measures required to enlist such broad-based support.

This chapter is organized into six major sections, each of which identifies challenges and costs that many districts typically neglect when completing their technology planning efforts.

1. Learning Resources not Included with the Boxes
2. Organizational Impacts & Management
3. Network Management & Development
4. Network Resources
5. Research & Development
6. Spirit & Support Building

The True Cost of Ownership

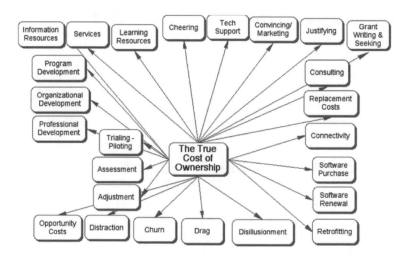

I. Learning Resources not Included with the Boxes

Despite the huge expenditures required to fill classrooms with networked computers, most schools discover late in the game that these shining boxes are relatively empty shells and the free Internet proves seriously lacking as an educational resource.

A. Subscription Information Resources

While some school leaders expect the Internet to provide a free lunch of quality information sources, the so-called "free" Internet is seriously flawed and unreliable. Schools discover that critical categories such as biography and history are not reliably and consistently covered. They invariably end up paying substantial fees to subscribe to periodical collections and other digital resources to supplement what comes free of charge. They also come to see digital resources as a poor substitute for a well designed print collection - the library. The costs of new digital resources mount while more traditional library expenses persist. In those cases where library collections are ignored, neglected, and underfunded, staff and students will find the search for truth severely compromised.

The True Cost of Ownership

B. Curriculum & Lesson Planning Services

Because the Internet is quite disorganized and often frustrating, many schools end up paying for curriculum support services like Classroom Connect and Scholastic or educational portals of various kinds such as BigChalk.Com. These services offer lesson plans and user interfaces that promise to convert the chaos of the Net into something comfortable, efficient and standards-based.

C. Supplemental Learning Resources

Many educators find themselves spending money on guidebooks, lesson packs, supplemental materials, and printed mater to accompany the new technologies. Despite vague promises of paperless classrooms and digital schooling, the new technologies often fall short of these visions and fail (by themselves) to provoke or sustain learning that is vivid, engaging, and effective.

Just as tourists wander through cities with printed guidebooks and maps, many visitors to the Net find a need for guidebooks and maps to make their visits productive.

II. Organizational Impacts and Management

It turns out that the health of the district, the effectiveness of the school and the productivity of the staff can be adversely influenced unless the introduction of the new program takes into account many organizational impacts such as the ones listed in this next section.

A. Assessment

In order to steer a program forward with good results, the team leading the innovation needs to collect data on a frequent basis. The data reveals which elements and strategies are working - which are failing. The data should cover everything from attitudes of staff and students to indicators of learning. If students are expected to work better in teams, the school should track development of those capabilities. If inferential reasoning and problem-solving are goals, measures of these capacities are critical. Those designing this assessment must be careful not to put too much focus on the wrong things, since the assessment tail may wag the program dog.

The True Cost of Ownership

While gathering data about the innovation, it also pays to monitor the climate of the school and district in more general terms. Indicators of general well-being and student productivity should be routinely checked to make sure the innovation is not having a negative impact elsewhere. In too many cases, the introduction of new technologies is approached as if such changes can only bring benefits. Stories of stress, turbulence, disappointment, and disillusionment would argue otherwise.

B. Adjustment

Programs almost always need to be adapted to match local conditions and circumstances. Based on assessment data and actual experience, some elements are eliminated while others are intensified. Fine-tuning is an essential aspect of success. During this process, denial is a major trap to be overcome, as leaders may have invested so heavily in the innovation that they wish to hear nothing but stories of grand success. They may be prone to shoot the messenger bringing tales of sorrow or disappointment.

Adjustment is unlikely to occur unless leaders with influence and good standing within the organization are charged with this responsibility, sheltered from risk and given healthy budgets to support the process. There must be a district or school planning process that institutionalizes and honors the invention process.

C. Opportunity Costs

Schools need to examine honestly what they will lose or pass up by implementing the new program. Perhaps the flow of funds and other resources into the new venture will starve existing projects or delay others that are deserving in their own right. If this starving and delaying is done consciously and intentionally, it is one thing, but if these consequences are unforeseen and ill-considered, the innovation may be guilty of distorting agendas and priorities - hogging the limelight and the lion's share of scarce resources.

As with the challenge of program adjustment, this kind of questioning and thinking is unlikely to occur in most places unless there is a formalized planning process and group that elevates risk assessment to a priority. Normal school operations are tinged with too much urgency and stress for personnel to address these issues during the course of daily management. The questions are only likely to surface in a productive and protective manner if they are handled within the deliberations of a planning group whose members have been trained to address doubts, risks, and challenges.

D. Churn

Teachers need some stability and predictability to meet the needs of young people. Too much change in the basic rhythms and routines of daily practice can throw teachers off balance and weaken their effectiveness. When the demands of the new new thing run counter to their prime mission - the nurturing of student performance - the broad range of goals is likely to suffer. Much as proponents of change enjoy metaphors of white-water rafting and turbulence, most classroom teachers require something quite different. Ironically, classrooms are more likely to change when there is a degree of stability.

E. Drag

New programs can place extraordinary demands upon the system and slow down progress in other areas, as the innovation requires a disproportionate share of resources just to get off the ground. Planners too often focus on the new without considering how it might impact what pre-existed. New gains may entail old losses, as the new new thing may inspire preoccupation and neglect.

F. Disillusionment

Because new programs almost always demand a high level of persuasion to convince folks to climb on board, there is a good chance that the reality will fail to meet the high expectations created by the hype and promotion. Broken promises and shattered dreams make for a long-standing reluctance to invest in future ventures.

G. Distraction

There is some danger that the innovation will sidetrack, derail, and delay the primary business of the school as everybody tries to launch the exceptional at the expense of the normal.

The True Cost of Ownership

III. Network Management & Development

A. Consulting

The complexity of the networking challenge puts many school districts in an awkward position. Because the innovation is relatively new, it is unlikely the district will possess strong internal resources to guide the process. One solution is to hire outsiders to work as district employees, but this solution is rarely sufficient, as the scale and complexity of the challenge is still likely to be grander than what can be handled by the type of person the district may be able to afford. Smart districts will maintain consulting relationships with outsiders that allow triangulation of advice - the comparing and contrasting of strategies without developing major dependencies. The trick is to create advisory relationships without conflicts of interest or dependencies.

B. Connectivity

The more that staff and students make use of a network, the greater the bandwidth required to support robust use. Unfortunately, many districts are penny wise and pound foolish when it comes to bandwidth, falling into the 11:00 A.M. trap. For some reason, student and staff use often peaks at late morning, resulting in such heavy traffic on the network that it is virtually impossible to reach any site on the Internet with acceptable speed. Many schools underfund the connectivity to such an extent that they end up with "virtual Internet."

C. Tech Support

In the business world, standards for technician support levels usually call for one technician for every 75 users. In school districts, there may be only one technician for every 300-500 desktops. This lack of staffing usually leads to an unstable network, a backlog of malfunctioning desktops and much frustration for end users. Understaffing also chokes off the development of many network services such as storage, e-mail and the provision of information services as crisis management (putting out fires) becomes the main focus of activity. Network starvation occurs when IT departments do not have enough staff to support robust network functions. In such cases, extensive use by teachers and students may actually compromise the performance of the network. Reduction or the blocking and discouragement of use may become a survival strategy.

IV. Network Resources

A. Software Purchase

In some respects, new technologies are empty shells, taking on power and value only as software programs are added to match the tasks that must be completed. Once the budget is expended for hardware, many districts find themselves without the funds to add a rich assortment of software tools to the computers. Many computers come bundled with basic software that must suffice for financial reasons. The software resides on the computers because it was bundled, not because it was chosen to match school objectives. To optimize educational use, a full array of curriculum-related software tools should be available for staff and student use.

B. Software Renewal

While software bundles may seem cheap, licenses often expire after a few years, requiring huge expenditures rarely anticipated or included in budgets. Networks change the debate over software piracy as vendors find it increasingly easy to monitor compliance and impose fines for violations. Maintaining current software licenses turns out to be costly.

C. Replacement Costs

Few districts have any real sense of the pace (and the price) of obsolescence. Advances and changes in network and system software mean that most computers cannot survive much longer than 36 months. Few replacement budgets match reality. This phenomenon either leads to rust or unanticipated spending lurches.

D. Retrofitting

Many of the existing infrastructures (both technical and human) will need to be upgraded to support the new systems and new experiences arriving upon the scene. Electrical capacity may be insufficient.

The True Cost of Ownership

Certain human capacities may need augmentation - collaboration, inventiveness, and flexibility. Too many assume that transplants may be executed without preparation.

V. Research & Development

A. Program Development

Given the highly unorganized and unreliable nature of the Internet, success requires the creation of learning experiences that match local curriculum and preferences. The district must either buy or create unit and lesson plans that convert the Net into something useful.

In order to see widespread use of these units, the district should be engaging teachers in both the construction and the testing of curriculum units, pilot-testing them to make sure they are worthy. Once they have proven worthwhile, they should be incorporated into district curriculum documents as required activities and should be supported by appropriate professional development programs designed to equip teachers with the skills they will need to implement the units effectively.

B. Professional Development

Few teachers are naturally equipped to make productive use of new technologies. They often require 50-100 hours of intensive adult learning to grasp the potential of new technologies to transform student learning. Very few districts are equipped to offer much more than software training, a trap that reduces the chance of making an impact upon daily classroom practice.

We now know that many teachers may require as much as 30-90 hours of curriculum-rich learning and unit building to achieve a high level of success with new technologies.

The design of successful adult learning programs is fully described in Chapter 15 (*).

C. Organizational Development

Some of the most promising change strategies require informal support networks that surround teachers with skilled and willing partners who can ease the path from disengagement to productive use.

The True Cost of Ownership

Creating a culture of support, encouragement, collaboration, and trust is no simple task.

D. Trialing - Piloting

In the rush to network rapidly, most school districts ignored a long-standing tradition of piloting major new programs in a small way before making a major commitment. Pilot programs allow decision-makers to figure out which innovations are likely to succeed and pay real returns on investments without making much of a commitment up front. It is a wise, discerning strategy that protects all from wild speculation and lurching.

In the next decade, as the exuberance of the dot com era passes, some of these time-tested strategies will regain popularity and return to common usage. Schools will be far less likely to adopt sweeping changes without seeing greater evidence of value up front.

VI. Spirit & Support Building

A. Cheering

Blending new programs and tools into appropriate daily classroom use involves considerable risk and stress on the part of teachers implementing the changes. To sustain such risk-taking and such innovative behavior, leaders and participants alike must engage in morale-building activities to keep everybody on the team feeling good about the change process. These activities require a substantial investment of time, energy and talent. Among other tasks, team members must monitor attitudes and moods frequently to make sure that bad feelings are kept from growing, spreading and contaminating the project.

B. Convincing/Marketing

All stakeholders must have a chance to consider and then applaud the wisdom of the innovation. To take any of this support for granted is dangerous. To assume that persuasion is not required is flirting with imposition and rejection. An energetic and full launch requires active, purposeful recruitment of all whose endorsement and approval will be required to achieve lift off. There should be no confusion or doubt about why the innovation is being introduced.

The True Cost of Ownership

C. Justifying

Once the district commits major resources to an innovation, a good deal of energy must go into the public relations effort required to convince all investors that it was a wise move. "It goes without saying," is a suicidal strategy, as skeptics and cheerleaders alike wish to know that their predictions proved true.

D. Grant Writing & Seeking

Funding innovation sometimes requires extensive investment in often frustrating grant-seeking behaviors as operating funds may not be adequate to support bold new ventures. Unfortunately, the grant seeking may be more hit than miss, requiring hundreds of hours of activity not easily translated into real benefits for classrooms and the program.

In other cases, as documented in a study by Means, et. al., successful grants may support only a short term burst of innovation rather than sustained program development. A sudden infusion of funds without long-term prospects for ongoing support may distort district priorities without making enduring changes.

Anticipating the True Cost of Ownership

Given the extensive list of costs outlined in this chapter, what can a school or district do to make sure the plan of action is complete?

The first step, of course, is to promote awareness across all stakeholder groups of these extensive costs and responsibilities. Unless they have been through such an implementation previously, many of these participants may be surprised by the contents of this chapter.

Once awareness of costs is achieved, the next step is to create a budget for the project that identifies all costs, whether they be costs that show up as dollar figures or costs that are felt in more subtle ways. For each of these costs there must be resources allocated or plans prepared to make sure costs do not become barriers or serious flaws in the adoption.

A district that employs a planning model similar to the one introduced at the beginning of this chapter will find that many of the costs outlined here will be addressed within the context of curriculum planning - a healthy and welcome approach.

Despite the pressure from vendors and their political friends to go entirely digital, there are still many times when paper may play a superior role supporting student investigations and problem-solving.

Schools must learn to protect themselves from digital orthodoxies - marketing statements designed to propel us into buying and using tools we may not need, may not be able to afford, and might find less effective than some of our time-honored non-digital friends such as Post-it® Notes, composition books, and white boards.

Digital classroom? Digital school? Digital generation?

These prescriptions are folly - an example of tools driving curriculum. Digital is automatically aligned with the notion of progress and improvement, even though history is littered with the wreckage of similar technologies that failed to please the public.

The dot com phenomenon should serve as evidence that not all innovations lead to improved conditions, operations and results.

Smaller is better?

At one point I overheard an Apple rep brag at a conference that a small laptop offered a "full half screen."

In the drive to offer lightweight, convenient tools, screen size (and lighting) becomes an issue, yet there is little open discussion of the limitations and frustrations associated with these new tools.

During a three day hands-on workshop this past year, I noticed that laptop equipped teachers kept turning to the printer to create paper documents that ended up on the desktop alongside books and the laptop. They were busy studying ways to address the needs of the Snake River (see workshop activities at "A River in Trouble" - http://fnopress.com/bigsnake/index.htm). As their Inspiration™ diagrams grew larger and more complex, they crept outside of the viewing range of the laptop screens. Six pages are hard to view on screen.

Wanting to see "the big picture," participants often printed out the six pages of the diagram and then taped those pages together in a large mind map.

In some cases, the mind map sprouted a flock of yellow sticky notes. Wisely, these teachers were supplementing the digital resources with print resources. The paper technologies sometimes proved superior to the digital in a number of respects. Viewability and flexibility were high on the list of features associated with paper.

At technology conferences, I often ask audiences if they print out articles found on the Web in hard copy in order to read them. The percentages that do such printing range between 70 and 90 per cent. Many of these folks are early adopters and technology enthusiasts, but they explain their printing in terms of convenience and the ease with which they can mark up, underline and interact with this older technology.

Since starting FNO Press four years ago, I have been astonished by the large number of books sold to educators who might find much of the same information and many of the same articles available for free online at **FNO**. There is something about a printed book that works very well for people even though many books can be downloaded as e-books from the Web.

The computer is fine for e-mail and browsing, but there is something about curling up on a comfortable chair with a book that laptops and handhelds simply fail to match.

The Myth of Supercession

Paul Duguid explains in "Material Matters" - an essay contained in **The Future of the Book** (Nunberg) - that many people foolishly assume that "complex new technologies will sweep aside their predecessors." Supercession, according to Duguid, is "the idea that each new technological type vanquishes or subsumes its predecessors: 'This will kill that.'" He provides examples of technologies such as hinges and pencils that have endured long past their predicted demises.

How does this relate to schools? Many technology cheerleaders promote new tools and digital schools without really knowing much about the consequences. In some cases, leaders call for laptops for all children in a school or grade or district or county or state as if such equipment will radically improve student learning.

It is actually possible to over-equip a school. There are times when digital is inferior. There are times when laptops should be shared and times when they should be laid aside.

Paper still works. Smart schools will resist thinly veiled marketing appeals to modernity that promote new tools without paying much attention to worth.

The new new thing (from **A Silicon Valley Story** by Michael Lewis) is not always the best thing.

:

Chapter 12 - The Medium is Not the Literacy

As noted in the previous chapter, the term "digital literacy" is often invoked by technology cheerleaders and promoters as part of their "rapid change" mantra. Things are moving so fast, the argument goes, that we need 21st. century skills and digital classrooms, digital schools, and digital frames of mind.

They would have us believe that what we did last year, the year before and during the 90s, is now outmoded. Time for a literacy upgrade! The new model. Literacy Version 2.0.

This pressure to adopt a digital life style, a digital pedagogy, and a digital mental set is often a thinly disguised marketing effort aimed at increasing sales of digital products and services at a time when telecommunications companies have overbuilt the bandwidth and are frustrated by low consumer demand and governmental investigations into their accounting practices. When value is lacking, they can always try to sell life style, fashion, and trend.

One network vendor at NECC 2002 was overheard pushing "rich media" – read "fat files" demanding big bandwidth and robust networking equipment. Most school people would rather think of rich media in terms of worthy and reliable content. Low fat, high value!

Now that the so-called "New Economy" and its dot com revolution have proven to be more speculation than reality, we turn to a so-called New Literacy. It pays to remember the fate of computer literacy.

I. What is Literacy?

There are more than a dozen literacies. Each refers to the challenge of wringing meaning and understanding from a type or source of information. We apply the term "information literacies" to subcategories of information:

artistic literacy	media literacy
ethical literacy	visual literacy
numerical literacy	text literacy
social literacy	cultural literacy

The Medium is Not the Literacy

We employ text literacy for understanding words and paragraphs of print whether appearing upon sheets of paper, upon computer screens, or upon the walls of ancient temples. We utilize numerical literacy when interpreting data collected on paper, in spreadsheets, on an abacus, or in online databases.

The focus of literacy should be upon understanding and interpreting various types of information. The tool on which a collection resides is secondary and incidental - unless one is intent on moving products. The tool becomes primary. Then we then witness classic cart before horse.

It trivializes the important concept of literacy to link the term to the driving and manipulation of equipment and formats. We should not suffer automobile literacy, bicycle literacy, paper literacy, analog literacy, computer literacy, digital literacy or fools.

2. Canvas Literacy?

At the San Francisco Fine Arts Museum, we can view a Winslow Homer oil painting of a fisherman threatened by an approaching storm – "Swell of the Ocean." Is that canvas literacy? Oil literacy? Hardly.

We can enjoy another version of the same painting printed out on paper and framed to hang in our home or classroom. Is that experience paper literacy? Hardly.

We can also view a scanned (digitized) version of this painting online at http://www.thinker.org. When we explore the metaphorical possibilities of this painting, we are employing visual literacy skills. For literacy (understanding) purposes, it does not matter much whether the painting is viewed digitally or as oil and canvas. Many of us might prefer the "real thing," but the type of literacy is determined by the type of thinking required, not the delivery system involved.

Despite the oft quoted "The medium is the message," (Marshall McLuhan), literacy is primarily about the message, not the medium used to serve it.

3. Analog Literacy?

The same could be said for media literacy. We read a news story about the FBI on a newspaper delivered to our front door in the morning. Maybe we check out the same story online at the **New York Times** Web site. We follow up by watching news coverage on **CNN** live (analog), and then we might turn to CNN.com to read text and view digitized video coverage.

The Medium is Not the Literacy

We combine text, visual, and media literacies as we try to make sense of this news story. We hope to understand how the words and images combine to represent reality, but we also apply critical thinking skills to consider the extent to which coverage of this story has been sensationalized. Does the **New York Times** coverage bring us closer to the truth than the **CNN** coverage? Does it matter whether it appeared online, on paper, or via satellite transmission?

Is the task of understanding news coverage substantially different just because the story appeared in digital format? Hardly.

The skills we apply when attempting to make sense of our worlds (information literacy) include analysis, interpretation, inference and synthesis. These have been a challenge and a struggle since the times of Socrates. They are not new skills.

The packaging and the digital spin are the main new ingredients.

4. Churn vs. Real Change

Proponents of new century skill lists pay little heed to the research on change in schools and other organizations. While we have little evidence that confrontation or criticism works as a keen motivator, the message that comes across with these models is a rejection of most teachers' stance, style, and practices.

Best practice is usually portrayed as some intensely wired learning experience with students passionately engaged in some digital environment. Unfortunately, this kind of digital utopia rarely matches its press releases.

We stand a better chance of recruiting the enthusiastic participation of traditional teachers if we show respect for their classical skills and strengths. When we set up false dichotomies (traditional=bad and constructivist=good), we run the risk of alienating the folks we need to join us.

Chapter 13 - Great Lesson Design

The Test of Monday Morning

"Can I use this on Monday morning?"

A great lesson will pass the test of Monday morning. It should have enough practical value to slide comfortably into the weekly classroom agenda without disruption, stress or drag upon the program.

We must take care when pressured by state leaders and others to "integrate technology" into our daily lessons because the process may be putting the cart before the horse. The very phrase implies some kind of injection from the outside that may not flow naturally out of healthy lesson design. We may end up injecting technology for the mere sake of technology, sacrificing student learning and sound lesson design principles on the technology altar.

Sound lesson and unit design will flow out of some greater sense of purpose than the integration of technology (note the singular) into a lesson. We might select a model such as Wiggins and McTighe, **Understanding by Design** and start unit development with a focus on the key concepts and understandings we wish to emphasize. A teacher or planning group then logically proceeds to identifying or creating learning experiences likely to support student learning of those essential questions and concepts. Strategies should flow naturally out of conceptual goals, and the selection of tools should follow clarification of purpose. We should put the program horse before the technology cart.

When proceeding with this type of unit design, teachers may find that digital resources are only appropriate for a few of these lessons and learning activities planned for several weeks of study. The teacher only adds digital lessons employing new technologies when such lessons will thereby be enhanced beyond other strategic choices. In those cases where library books, interviews or field study may be preferable, they should prevail.

Great Lesson Design

Module Maker 2

An example of this kind of lesson design can be found at http://questioning.org/module2/quick.html. **Module Maker 2** was invented to support teachers in creating digital lessons in ways that might comfortably emerge as part of larger unit. Excerpts are reproduced here, but online examples will be found at the above URL.

Can we invent curriculum-rich lessons that take students half an hour but engage them in powerful thought with considerable skill?

Module Maker 2 equips teachers to create standards-based learning experiences that combine good print information with good digital resources.

In 30 minutes, a team of students can wrestle with a big question and cast some light on it, whether they be first graders, pre-schoolers or approaching middle school.

The secret is good, tight lesson design.

No waste, no bother and no wandering about.

Structure, clarity, and meaning!

We need several elements to build a great lesson:

1. An intriguing question that matches the standards.
2. A collection of information that will spark understanding.
3. Thought-provoking activities.

Provoking a sense of wonder is paramount.

Curiosity rarely kills a cat, but boredom is definitely lethal!

The 3 Prime Questions

Certain questions provoke more curiosity and wonder than others. Three stand out as especially powerful:

Why?
How?
Which is best?

1. Why

Why do things happen the way they do?

This question requires analysis of cause-and-effect and the relationship between variables. It leads naturally to problem-solving (the "How?" question) or to decision-making (the "Which is best?" question).

"Why?" is the favorite question of four-year-olds. It is the basic tool for figuring stuff out (constructivist learning). At one point while researching student questions in one school district, I found "Why?" occurred most often in kindergarten classrooms and least often in the high school (which had the highest SAT scores in the state).

Why does the sun fall each day?

Why does the rain fall?

Why do some people throw garbage out their car windows?

Why do some people steal?

Why do some people treat their children badly?

Why can't I ask more questions in school?

2. How?

How could things be made better?

This question is the basis for problem-solving and synthesis. Using questions to pull and change things around until a new, better version emerges.

"How?" is the inventor's favorite question.

"How?" is the tool that fixes the broken furnace and changes the way we get cash from a bank.

"How?" inspires the software folks to keep sending us upgrades and hardware folks to create faster chips.

"How?" is the question that enables the suitor to capture his or her lover's heart.

"How?" is the reformer's passion and the hero's faith.

3. Which is best?

Which do I select?

This question requires thoughtful decision-making - a reasoned choice based upon explicit (clearly stated) criteria and evidence.

"Which?" is the most important question of all because it determines who we become.

Which school or trade will I pick for myself? Which path will I follow?

Great Lesson Design

Two roads diverged in a yellow wood,
And sorry I could not travel both
And be one traveler, long I stood

Robert Frost

Faced with a moral dilemma, "Which path will I follow?" Confronted by a serious illness, "Which treatment will I choose for myself?"

Mining the Standards

We must learn to pull these intriguing questions out of state or provincial curriculum standards . . . questions that require analysis, interpretation, inference and synthesis even from our youngest students.
What can we find below the surface?
A question must pass the test of "So what?"

"What did you do in school today?"

"Nuthin."

Even very young children can tackle challenging questions if we frame them carefully in words that make sense to them. First graders can compare and contrast fast food restaurants, for example. Four-year-olds can talk about ways to help a sad friend.
Even young children can analyze, interpret, infer and synthesize.
Many states have issued demanding goals. Witness these science standards from Wisconsin:

Wisconsin
Performance Standards

By the end of grade four, students will:

C.4.1 Use the vocabulary of the unifying themes* to ask questions about objects, organisms, and events being studied
C.4.2 Use the science content being learned to ask questions, plan investigations*, make observations*, make predictions*, and offer explanations*

Great Lesson Design

C.4.3 Select multiple sources of information to help answer questions selected for classroom investigations*
C.4.4 Use simple science equipment safely and effectively, including rulers, balances, graduated cylinders, hand lenses, thermometers, and computers, to collect data relevant to questions and investigations*
C.4.5 Use data they have collected to develop explanations* and answer questions generated by investigations*
C.4.6 Communicate the results of their investigations* in ways their audiences will understand by using charts, graphs, drawings, written descriptions, and various other means, to display their answers
C.4.7 Support their conclusions with logical arguments

Finding Good Content

Digital information sources can be just wonderful, but they are often quite limited for certain subjects - especially when planning for younger children. It is hard to find pages designed with easy reading, large print, and attractive graphics.

Rather than start with great questions drawn from the standards, teachers might wisely start by prospecting for Web sites that have content matching the content for social studies or science in their curriculum.

Once they find good content, they can extract the analysis, inference, interpretation, and synthesis questions that will form the basis for the student learning activity.

Keeping Digital Lessons in Perspective

Most of the time we expect that students will learn about science and social studies through a mix of media and sources . . . some digital, some print, and some from nature and the community. The digital lessons we design are brief "chunks" that must fit into a larger web of activities.

If students are going to learn about storms, for example, perhaps they should begin with a brightly illustrated text or book from the library.

They might interview parents or grandparents about storm experiences.

We would add digital learning to our learning web only if we found a storm site that brought extra meaning and value to the study.

Great Lesson Design

Perhaps NOAA or some other agency would be a good source of information not readily available in print?

Prospecting for Digital Riches

Sadly, search engines are rarely a good tool to employ when looking for learning materials suited to young children. You can waste dozens of hours wandering around through hundreds of sites that are poorly designed, inaccurate and developmentally inappropriate.

It makes better sense to check out lists of content sites suggested by educators who have some judgment (and time) to separate treasures from trash.

One example is Kathy Schrock's **Guide for Educators**:
http://school.discovery.com/schrockguide

Another example is **KidsClick** - a site created by school librarians who have identified child appropriate material by subject heading.
http://sunsite.berkeley.edu/KidsClick!/

Relying Upon Templates

While templates can reduce the flair and creativity of online lessons, they do have the advantage of speeding the invention process. Rather than putting most of their time and effort into color, appearance and special effects, templates enable teachers to create lessons with little fuss or bother.

Those with HTML design skills may return to these lessons at a later date to modify design elements in line with personal preferences.

To create lessons using **Module Maker 2**, teachers will add words and graphics to 4-6 pages that will be used by students and other teachers.

These templates will be found online at http://questioning.org/module2/quick10.html

One Page for Each Key Element

1. The Essential Question and Learning Task
2. The Information Source
3. The Student Activity
4. The Assessment Activity
5. Enrichment Activities
6. Teacher Support Materials

Great Lesson Design

A sample lesson is available at http://questioning.org/module2/quick9.html

1. The Question

When you compare hurricanes during the past decade, which two deserve a place on the top of the list for damage and destruction?

To answer this question, students must compare the damage done by several hurricanes and rank the list from most damaging to the least damaging.

2. The Information Sources

Students will visit a government Web site to learn about the damage done by recent hurricanes - the National Climatic Data Center at http://lwf.ncdc.noaa.gov/oa/ncdc.html

* Hurricane Fran
* Hurricane Mitch
* Hurricane Georges
* Hurricane Bonnie
* Hurricane Bret

3. The Student Activity

It is the student's job to fill in the grid below, comparing the hurricanes. The teacher will either print the grid on paper for stduents, or they can build one as a table in their word processing program. Once they have all the information, then they will rank the hurricanes from most damaging to least damaging.

1 = top damage
5 = least damage.

Example on the next page . . .

3. The Student Activity

It is your job to fill in the grid below, comparing the hurricanes. Your teacher will either print the grid on paper for you, or you can build one as a table in your word processing program. Once you have all the information, then you may rank the hurricanes from most damage to least. 1 = top damage 5 = least damage.

Public domain from NOAA

Name of Hurricane	Damage $	Deaths	Wind	Other	Date	Rank
Hurricane Fran						
Hurricane Mitch						
Hurricane Georges						
Hurricane Bonnie						
Hurricane Bret						

The underlined names of hurricanes above link directly to information-rich pages at NOAA that will speed students toward insight and understanding.

A well scaffolded lesson offers a number of real advantages:

1) Provides clear directions
2) Clarifies purpose
3) Keeps students on task
4) Offers assessment to clarify expectations
5) Points students to worthy sources
6) Reduces uncertainty, surprise, and disappointment
7) Delivers efficiency
8) Creates momentum

4. The Assessment Activity

Finally, stduents write a paragaph telling which hurricane they would pick as the "storm of the century." They must provide convincing facts as evidence to back their choices.

Sometimes they ride into town on white horses. "We are about to become the first laptop school in the nation (or county or state)!"

Next thing you know, life is full of change and churn - not all for the good. What's wrong with this picture?

So many of these people specialize in creating anxiety.

"You're falling behind. Run faster! Close the gap!"

We must remember that churn is usually not good change. It has the roar, the foam, and the excitement of surf but is often "full of sound and fury signifying nothing." (**Macbeth**)

In contrast with the leadership style described in Michael Fullan's book **Leading in a Culture of Change**, visionaries take pride in the boldness of their ideas. They are happiest when skirting the bleeding edge of change, pushing the school or organization into uncharted areas.

Visionaries are supposed to have the capacity to see over the horizon - a talent not well distributed among mere mortals. All too often, self-proclaimed visionaries lack both the talent and the self awareness to move more carefully. Arrogance too often tags along-side of the vision.

One superintendent speaking at a national conference mentioned that he expected all teachers in his district to be rabbits (rather than snails).

"We create a zone of discomfort," he bragged. "Feed the rabbits. Starve the snails."

Quality, rigor, communication, and value may be left behind in the race to change.

In a time of rapid change and heavy marketing, schools are especially vulnerable to leaders who might fashion partnerships with technology companies without carefully considering the consequences.

Sadly, many visionaries are better at imagining the benefits of innovations (and giving speeches) than they are at managing the change process so that benefits are actualized. It is enough, evidently, that such leaders be able to grasp new strategies that will put the school or the district into some kind of leading position.

Beware the Visionary

Being early sometimes seems more important than getting it right.

As an example, a leader might promote the purchase of laptops for all students without understanding the organizational development or the infrastructure required to convert such an investment into a robust learning experience. This leader might ignore alternatives.

Without full engagement of the staff in program and professional development, conservative, skeptical staff members might pay little attention to the laptops and continue with business as usual. Without adequate investment in network access to the Internet, students who attempt Internet research might find themselves crawling at unacceptable speeds.

In times such as these, when we have seen dot com companies go bust, Enron accounting practices distort reality, and many bold new product ideas fail to deliver value, we need leaders capable of practicing discernment.

Churn often represents the opposite of healthy change. We sometimes watch a visionary leader turn everything upside down and inside out, speaking of a bold new future that makes little mention of classical values and traditions. They give the impression that the "new new thing" is all that counts. Staff is soon divided into true believers and skeptics. Discernment and questioning is viewed as recalcitrance and heresy.

So-called visionaries have come up with such disturbing and foolish notions as the digital library, suggesting that the Internet is an adequate source for most inquiry. Evidence to the contrary rarely penetrates their digitally narrow world view.

> Churn   VERB: 1. To cause to move to and fro violently: agitate, convulse, rock, shake. See CALM, REPETITION. 2. To be in a state of emotional or mental turmoil: boil, bubble, burn, ferment, seethe, simmer, smolder. See CALM.
> **Roget's II: The New Thesaurus**, Third Edition. 1995.

For a wonderful visual showing churn, note the Michael Leunig cartoon, "Vasco Pajama in the Strait of a Thousand Lighthouses" on the opposite page.

In the 1970s, this type of school leader was briefly heralded as a "change agent," but the term rapidly fell into disrepute as folks noticed that such leaders often moved from job to job without actually completing the work they launched.

Several decades of research into effective change strategies for schools have discredited this kind of leadership, suggesting instead, that sustained, deeply rooted improvements require much more than bold ideas. They require an intimate knowledge of local conditions, traditions, norms and resources. They demand sensitive, handcrafted orchestration of efforts. They cannot succeed unless the leader and her/his lieutenants do more listening and engaging than preaching.

Evangelism, strangely, is heralded by some technology advocates as a way to develop enthusiasm for new tools, but the judgment and fervor trailing such strategies often provoke opposition, resentment, and resistance rather than conversion.

"Take it on faith!" Not likely now that the bubble has burst.

© Michael Leunig, reprinted with permission.

Lighthouse Schools

Schools with an avid appetite for change sometimes refer to themselves as "lighthouse schools" or "lighthouse districts." As new strategies, trends, and notions of educational improvement come along, such schools take pride in adopting the innovations early on. They become famous for setting trends and embracing innovations early.

Beware the Visionary

Unfortunately, most untested new approaches will expose the lighthouse districts and schools to surprises and disappointments - what Fullan calls, "implementation dips." Smart leaders devote plenty of attention to planning for such surprises.

"What's the worst that could happen?"

A basic planning strategy, *Force Field Analysis*, calls for the listing of obstacles and issues the team might encounter prior to project launch. The team then creates an action plan or solution for each problem on the list.

Often, visionaries leave out this step. They have a tendency, like the mythical figure, Icarus, to fly too close to the sun. Their wax melts, the feathers fall away and they tumble, as he did, into the sea of disenchantment.

"What could possibly go wrong?"

And even when the implementation experience forces issues and problems out into the open, visionaries have been known to shoot messengers or wrap themselves in cloaks of denial.

"How are things going out there?"
"Great, boss. It's just like you said it would be."

A disturbing tendency with visionaries is the hiring of disciples - true believers who share the vision (and blindness) of the great leader and help maintain several layers of insulation from the news of issues and problems. Those who share alarming data or dare to question the vision, are often branded heretics and exiled in ways that will prevent them from gaining a broad audience.

During this decade of rethinking the value of new technologies in schools, we need leaders who prize discernment, pacing, well-being, and focus. We've had enough bandwagons, silver bullets, and quick fixes to last a long time.

Chapter 15 - After Laptop

What is the logical next step for a school that invested heavily in laptops five years ago but is having second thoughts?

Are there reasonable alternatives to each student owning a laptop?

Is there a network design and delivery system that would create more good?

Could sharing be one answer?

Can schools free parents from the costs of keeping up with the rapid obsolescence so typical of laptops?

> ". . . as a general rule, the longer a student was involved with the laptop project, the less the likelihood that the laptop was used for work at school."
>
> Quoting (out of context) from the third year evaluation report of a laptop program in Beaufort County, SC.

Some schools jumped on the laptop strategy in a big way five or ten years ago by requiring all parents to buy a laptop for every student.

Seemed like a good idea at the time.

But that's a lot of money, especially for a family with 3-4 children. And not everything worked out exactly as planned in some of these schools.

Sometimes the schools had not fully planned how the laptops would be used. Sometimes they forgot to ask staff if they thought this was a good idea. Sometimes they failed to set aside much money for program or professional development.

Sometimes some of the teachers went for long periods of time without asking students to turn on their laptops.

After Laptop

Sometimes some of the students left laptops at home.

Sometimes the parents asked why they had invested so much money in tools that were used sporadically.

Sometimes teachers argued that new digital resources were often unreliable and of poor quality.

Sometimes bandwidth was so poor at the school, students had "virtual Internet."

Sometimes batteries would only last 90 minutes.

A Closely Held Secret

A decade after the creation of so-called laptop schools (ones where each student owns or carries an individual laptop), it is difficult to find anything but glowing reports and testimonials regarding the benefits of equipping schools in this manner.

Visit some of those schools and speak with rank-and-file teachers and the pictures projected to the outside world sometimes contrast with the images that emerge from within. At times it feels like "The Emperor's New Clothes." It is OK to whisper off the record that there have been serious issues and disappointments, but these issues and disappointments remain, typically, a closely held secret.

This silence is unfortunate because it keeps successive waves of planners in the dark by depriving them of the information that might help them launch more effective programs or select quite different strategies.

Three Years of Data from One Project

The one major exception to this "dark hole" is a series of three annual reports conducted by the Beaufort County Public Schools, a district that made a major commitment to a laptop program and presented at national conferences advocating the benefits of all students having laptops.

To the district's credit, a substantial commitment was made to program evaluation and data gathering - a rare thing in the technology business. Again to the district's credit, the results are published for public review so that others may learn from the data.

Year Two Study
http://www.beaufort.k12.sc.us/district/ltopeval.html

After Laptop

Year Three Study
http://www.beaufort.k12.sc.us/district/evalreport3.htm

It was revealing to read through the early public relations documents and then review the evaluation reports over the span of the project. High hopes and declarations of achievement faded in the face of data less impressive than early hopes.

Stevenson's reports are thorough and impressive, taking many pages to review the three year experience. There are many issues and elements of these reports that deserve close reading by any school thinking of moving toward a full laptop program. At one point, for example, he reports on the major barriers perceived by teachers as blocking effective implementation of the program.

It seems evident that many teachers gave this program a serious and earnest effort but that many factors created frustration.

The Laptop Promise

Several years ago, after attending a national conference promoting laptops for all students, this author published an article in **FNO**, "The New New School Thing," critiquing the exaggerated claims of technology cheerleaders. The following table was included. (see http://fno.org/apr2000/newnew.html)

If you buy a laptop for each student, proponents promise . . .
• Better writing
• Expanded knowledge
• Increased achievement - higher scores
• Improved skills for the modern workplace
• Enhanced learning and teaching efficiency
• Heightened motivation for all involved
• Enriched preparation for global citizenry
• Elevated problem-solving and decision-making
• Intensified student-centered learning
• Augmented teaming and cooperation

Like many dot com companies, the proponents provided little or no evidence to back these claims.

Despite the claims of many laptop school proponents that the equipment will engender a different kind of classroom and different kind of learning, teachers in the South Carolina project listed note-taking, homework assignments, writing, electronic learning activities, and accessing the Internet as top choices, with cooperative learning, student research, and student presentations falling much lower in the list.

That a very large county effort has reported disappointing results after four years of working with laptops is a reality that needs to be publicized more broadly so that others can learn from the experience. The negative findings have not risen to the level of general public awareness, however.

As just one example, after three years of using laptops in class, participating eighth grade students were far from enthusiastic. Both students and teachers reported levels and types of use that are quite disappointing.

Student Evaluation of the Impact of the Laptop Project on Academic Achievement - from Table 14 of Year 3 Evaluation Report

13 Per Cent - Per Cent First Year Students Indicating Laptops Having "Positive" Impact

14 Per Cent - Per Cent Second Year Students Indicating Laptops Having "Positive" Impact

16 Per Cent - Per Cent Third Year Students Indicating Laptops Having "Positive" Impact Impact of the laptops on academic success in 1998/99

How Much Students Used Computers for School Work "At School" in 1998/99 - from Table 6 of Year 3 Evaluation Report Response

1. How Often First Year Students Were Using Computer at School for School Work

    Never   6 Per Cent
    Some   40 Per Cent
    A lot   52 Per Cent

2. How Often Second Year Students Were Using Computer at School for School Work

    Never   7 Per Cent
    Some   52 Per Cent
    A lot   40 Per Cent

3. How Often Third Year Students Were Using Computer at School for School Work

    Never   26 Per Cent
    Some   45 Per Cent
    A lot   27 Per Cent

Teacher Evaluation of the Impact of the Laptop Project on Academic Achievement - from Table 5 of Year 3 Evaluation Report Topic

40 Per Cent - Per Cent First Year Teachers Indicating Laptops Having "Positive" Impact
46 Per Cent - Per Cent Second Year Teachers Indicating Laptops Having "Positive" Impact
52 Per Cent - Per Cent Third Year Teachers Indicating Laptops Having "Positive"
Impact Impact of the laptops on academic success in 1998/99

Conclusion

Before launching innovations for entire districts and states, responsible leaders should review the data on previous attempts and give serious consideration to launching several different pilot programs that allow for comparisons and adjustments.

There are at least 5-6 different ways to equip a school program with computing resources. One model is to buy a laptop for every student and teacher - leaving little funding for program and professional development. Another model is to buy laptops in waves, moving towards the purchase of large numbers only as teacher readiness and inclination is fully developed across the entire faculty.

A third model is to buy laptops and move them around on carts as proposed in this book.

It makes sense to move deliberately and with moderation. This is a time for caution rather than plunging forward putting carts before horses.

Chapter 16 - Stories of Adult Learning

Without investing heavily in adult learning and program development to set the stage for frequent and effective use of new technologies, it is foolish to jump ahead to the installation of networked computers throughout a school.

This chapter outlines the characteristics of standards-based, worthy uses of new technologies. It then describes the characteristics of an effective district change process - one that engages all teachers in an inviting and generative adult learning journey. The key term is "generative" - meaning that behaviors and daily practice will be changed for the better as a consequence of professional development experiences. Such change does not result from simple software training. The adult learning must be curriculum rich and clearly focused on enhancing student performance.

1. Disillusionment

In the past few years, schools have spent fortunes wiring classrooms in the hope that such investments would transform learning and produce gains in student performance. As this book goes to press, there is little credible evidence to justify such expectations.

Sadly, much of the "digital revolution" urged on schools has proceeded without noting the research describing how teachers learn challenging new strategies (Joyce, 1990; Leiberman, 1995, 1999). Those intent on wiring schools usually ignored the literature outlining the kinds of organizational development required to make successful changes in schools (Fullan, 1991; Lieberman, 1995, 1999). In most cases, they even ignored warnings from the world of business about "late adopters" (Moore, 1991) and "the total cost of ownership" (Van Dam, 1999).

Trouble begins when folks put carts before horses.

Wisely, it seems, many teachers resist powerpointlessness - a term coined by one Australian educator - the flashy, glib use of technology for technology's sake. According to Becker (1999), traditional teachers (more than 60% of his survey) seek classroom activities that will help students perform well on increasingly demanding state tests and curriculum standards. He reports that traditional teachers are three times less likely than their constructivist colleagues to let students use computers even when they have five PCs in their rooms.

Data reported in **Education Week's Technology Counts '99** showed that teachers were not making powerful, curriculum rich use of their networks. They also found that most teachers reported that they were not well prepared to use new technologies.

"And a new **Education Week** survey has found that the typical teacher still mostly dabbles in digital content, using it as an optional ingredient to the meat and potatoes of instruction." (Trotter, 1999)

Unfortunately, most districts have provided very little professional development, with Market Data Retrieval reporting in 1999 (p. 122) that sixty one per cent of American teachers received 0-5 hours of technology related "training" annually. And much of the focus of such training has been on learning software (how to make slide presentations or use spreadsheets) rather than on curriculum blending and classroom strategies.

Sadly, comparable data for 2000-2001 is unavailable since funding for these reports disappeared with the curious disbanding of the Milken Exchange and its smart funding of such studies. **Education Week** did not produce a **Technology Counts 2000**, and the focus of **Technology Counts 2001** turned to "The Digital Divide." http://www.edweek.org/sreports/tc01/ - a different set of issues and questions.

It is unfortunate that groups cheering the advent of new technologies stopped gathering data that helped us to assess whether or not the investments were paying dividends and whether teachers were getting the professional development support they need. Even the CEO Forum, which stressed the importance of funding professional development in its 1999 report, has stopped assessing and reporting progress on such issues. The CEO Forum suggested with its **Star Chart** that penetration (a large number of computers installed per classroom) translates into curriculum integration. This claim has been dispelled by Becker's study linking frequency and types of use with teacher attitudes and readiness as well as the level of district investment in support services as opposed to hardware.

Stories of Adult Learning

II. Clarifying Purpose - Engaged Learning and Student Growth

In some schools, leaders ignored several crucial steps: 1) they forgot to clarify purpose and 2) they made little effort to win over reluctant and skeptical teachers. They jumped to the installation phase without dealing with the issues of inclination, philosophy, readiness, and support identified by Becker's study. It should be stated, in fairness, that some of the grant and funding programs gave them little choice.

"Install and they will log on!" was the apparent campaign slogan.

In order to enlist the enthusiastic participation of all teachers, our goal should be the strengthening of student performance on reading, reasoning, problem-solving and related tasks drawn from state curriculum standards (Note Virginia Standards below). It should not be the creation of digital classrooms and wired schools. New technologies should be used when they enhance learning. They should sit comfortably alongside older technologies such as books and paper. They should take a back seat when other modes of learning excel.

Virginia Standards of Learning (SOLs) - Science

"Investigate" refers to scientific methodology and implies systematic use of the following inquiry skills:

- Observing
- Communicating
- Measuring
- Predicting
- Inferring
- Hypothesizing
- Classifying and sequencing
- Designing, constructing, and interpreting models
- Interpreting, analyzing, and evaluating data
- Defining, controlling, and manipulating variables in experimentation

Visit a classroom making wise use of technologies and you will see a judicious blending of tools - no digital imperatives warping values.

Picture a middle school teacher in Virginia who has launched a major science unit challenging students to invent better ways to protect and restore various endangered species.

A cart of 15 wireless, Internet-connected laptops lives in the classroom for two weeks, providing "critical mass" - enough units to allow pairs and trios to share laptops to conduct their research and their writing. Thirty students spend their time fully engaged (see inset on Engaged Learning) with a variety of resources, some print, some digital.

The visitor notes that cluster diagrams mapping out research questions (using Inspiration™ software) have grown huge. Because they are too large for the computers' limited screens, some students have printed their diagrams out onto six or more pages of paper to spread them out across their desks. That way they can see the entire "map" at one time.

Characteristics of Engaged Learning

When learning activities have been well organized around intriguing issues and questions, the visitor notes that students are . . .

1. Responsible for their own learning - They invest personally in the quest for knowledge and understanding, in part because the questions or issues being investigated are drawn from their own curiosity about the world. Projects are pertinent and questions are essential.

2. Energized by learning - They feel excited, intrigued and motivated to solve the puzzles, make new answers and reach insight. Their work feels both important and worthwhile.

3. Strategic - They make thoughtful choices from a toolkit of strategies, carefully weighing which approach, which source and which technique may work best to resolve a particular information challenge.

4. Collaborative - They work with others in a coordinated manner, splitting up the work according to a plan and sharing good ideas during the search for understanding.

(Adapted from the **Engaged Learning** model - Jones, et al, 1994; Means, 1997)

The visitor also notes some teams pointing back and forth between pages of books and the screens of laptops as well as topographical maps showing the locations of their species. Multiple sources of information are alive and well.

Some students are reading from screens together. Some are writing and taking notes on laptops. Others are using sticky notes. Some are e-mailing to other students or scientists in states or countries where their species is threatened.

Several teams leave the room for the school media center, having found excellent books through the library catalog now available online in their classroom.

One team is excused for the day to interview and tape scientists at a nearby preserve. They have borrowed the school's digital video camera.

The teacher moves about the room, lending a hand, offering a nudge, and patting a back as needed, managing the flow of activities and the well focussed hubbub of the classroom with calm and comfort.

III. Setting the Stage for Success

Scenes like the one above do not happen overnight. They result from substantial planning and investment in teacher growth over several years. A robust program of professional and program development engages each teacher in 20-60 hours of learning and inventing annually, with the U.S. Education Department urging a commitment of 30 per cent of any technology budget to such activities.

"Today, schools spend an average of 9 percent of their technology budgets on training and support, while the experience of technology-rich schools suggests that more than 30 percent of much larger technology budgets should be invested in these areas." (U.S. Department of Education, 1996)

The development of the endangered species unit described above was the culmination of several years of learning and growing. The teacher began the journey with many hours of actually using the new tools for research and then moved on to many more hours of study group participation before finally attempting to build and teach a technology enriched unit.

The teacher's unit planning took place in the summer of 2001 as a team member participating in a week of district sponsored (paid) curriculum development. Working with two other science teachers and a media specialist, the team first identified standards-related essential questions and concepts with an approach suggested by **Understanding by Design** (Wiggins and McTighe, 1998). Once the learning goals were clear, the team sought a combination of learning activities that were likely to produce the desired outcomes.

Even before the summer invention project began, this team had worked as a study group through the 2000-2001 school year exploring promising practices such as **Understanding by Design** and **Power Learning** http://fno.org/PL/powerlearn.htm.

The summer before that year of study, in August of 2000, they had all participated in three days of hands-on problem solving that engaged them in exploring an environmental issue with the new information technologies (http://fnopress.com/bigsnake/index.htm). In "A River in Trouble," they were charged with making recommendations to the

President with regard to the issues facing the Snake River.

Finally, in the summer of 2001, they were able to translate these theories and these learning experiences into the building of units for the 2001-2002 school year.

IV. Invention as Professional Development

While it is rarely recognized as such, the invention of effective unit plans can be a remarkably effective way to develop skills while winning support for the use of new technologies. The team mentioned above dove into unit planning with tremendous enthusiasm.

Rather than condemning students to hours of fruitless wandering across a virtual jungle of Internet resources (Google offers 595,000 Web pages with the phrase "endangered species"), the team spent many hours identifying reliable sources.

By stressing quality from the outset, they addressed a chief complaint of many teachers in the school - a sense that the Internet was badly organized, unreliable and quite weak when it came to supporting standards-based learning.

The media specialist began by pointing out the middle school's excellent print collection of recent books treating the challenge of restoration and protection - books such as **A Common Fate: Endangered Salmon and the People of the Pacific Northwest** by Joseph Cone. It became evident as they compared digital and print resources, that some print sources might provide greater depth and value than the articles available on the Internet or within the district's collection of electronic periodicals. The team devoted more than a day to the development of a balanced and substantial list. As a result, they were able to set aside concerns about quality and felt optimistic that they might win over their more skeptical colleagues across the district.

The librarian also pointed the team toward some helpful Web sites such as a listing of Research Modules and WebQuests that provide highly scaffolded (structured) projects - http://fno.org/url.html. The team decided to adapt and adopt an endangered species unit from Grand Prairie, Texas that seemed well organized and quite promising - http://www.gpisd.org/gpisd/modules/Highschools/GPHS/GPToad/VisitOne.html.

The librarian also suggested rresources available from the Environmental Protection Agency at http://www.epa.gov/students/ecosyste.htm as well as lists of good resources compiled by other educators such as **KidsClick** - http://www.kidsclick.org/ and **Kathy Schrock's Guide** - http://school.discovery.com/schrockguide.

Stories of Adult Learning

The team welcomed this district-sponsored opportunity to provide focus, direction and quality to their students, knowing that class time would be spent more efficiently thanks to the preparation and planning made possible during the summer months.

V. Replacing Training with Adult Learning

How does adult learning differ from the training models that have dominated technology-related professional development for the past two decades?

Adult learning usually involves the learner in activities that match that person's preferences, interests, needs, style, and developmental readiness.

The learner makes choices from a rich and varied menu of learning experiences and possibilities but must take responsibility for planning, acting and growing.

If we shift school cultures to support adult learning, professional development is experienced as **a personal journey of growth and discovery** that engages the learner on a daily basis. In the best cases, adult learning includes an emphasis upon self-direction, transformation and experience. One learns by doing and exploring . . . by trying, by failing, by changing and adapting strategies and by overcoming obstacles after many trials. One learns by teaming - sharing failures and successes as well as tricks and techniques that work.

This approach to supporting teachers may actually generate a change in how classroom learning occurs.

Unlike the training models, adult learning is primarily concerned with creating the conditions, as well as the inclination and the competencies to transfer new tools and skills into daily practice. While training usually occurs outside of context and frequently ignores issues of transfer, adult learning is all about melding practice with context. Adult learning should encourage teachers (and their allies) to identify and then remove obstacles.

VI. Examples of Effective Practices

The menu for teacher learning expands dramatically from lists of classes offering training in software to many other options such as study groups and invention teams similar to those described earlier in this article. Each teacher creates a professional growth plan (PGP) outlining steps to be taken on a personal journey of growth.

To support such journeys, the district invests in support systems such as technology coaches, mentors, and cadres. Augmenting these formal systems, the district strives for "just in time support" by encouraging the development of informal support groups and support staffing of various kinds. Teachers have help lines and answers to frequently asked questions (FAQs) within easy reach. To expand their horizons and appetites, they find many opportunities to enjoy excursions such as school visits, work place visits, conferences, etc. In order to match lifestyles and provide 24/7 adult learning, districts explore the potential of online learning programs. Many of these offerings are described at length in "Head of the Class: How Teachers Learn Technology Best." **Electronic School**, January, 2001. http://www.electronic-school.com/2001/01/0101f2.html and **in How Teachers Learn Technology Best** (McKenzie, 1999).

VII. Measuring Return on Investment

There is far too little assessment being done to guide professional development. Most districts do not know the level of development already achieved by staff, let alone their preferences, styles, fears and passions. A thoughtful assessment strategy helps to identify offerings that stand a chance of matching preferences, and then assessment makes it possible to steer the program forward. (see "Finding Your Way through The Data Smog" by Joe Slowinski at http://fno.org/sept00/data.html.)

VIII. Putting New Technologies to Good Use

We have known for a very long time that enduring and worthy change in schools depends upon strategic and robust investments in professional and program development. If we expect to see new technologies employed in ways that make a difference in how students learn and think, it is time to put far more resources into adult learning.

References

Becker, Henry. **Internet Use by Teachers**. 1999. http://www.crito.uci.edu/TLC/FINDINGS/internet-use/startpage.htm
The CEO Forum Year 2 Report. 1991. http://www.ceoforum.org
Cone, Joseph. **A Common Fate : Endangered Salmon and the People of the Pacific Northwest**. New York: Henry Holt & Com-

pany, 1995.

Deal, Terrence E. and Peterson, Kent D. **Shaping School Culture : The Heart of Leadership.** San Francisco: Jossey-Bass Publishers, 1998.

Fullan, Michael G. **The New Meaning of Educational Change**. New York, Teachers College Press, 1991.

Jones, B., Valdez, G., Nowakowski, J., & Rasmussen, C. Designing Learning and Technology for Educational Reform. Oak Brook, IL: North Central Regional Educational Laboratory, 1994. http://www.ncrel.org/sdrs/engaged.htm

Joyce, B. (Ed). **Changing School Culture through Staff Development**. Alexandria, VA: ASCD, 1990.

Lieberman, Ann and Miller, Lynne. **Teachers—Transforming Their World and Their Work**. New York: Teachers College Press, 1999.

Lieberman, A. **The Work of Restructuring Schools: Building from the Ground Up**. New York: Teachers College Press, 1995.

McKenzie, Jamie. **Beyond Technology: Questioning, Research and the Information Literate School**. Bellingham, WA: FNO Press, 2000.

McKenzie, Jamie. "Head of the Class: How Teachers Learn Technology Best." **Electronic School**, January, 2001. http://www.electronic-school.com/2001/01/0101f2.html

McKenzie, Jamie. **How Teachers Learn Technology Best**. Bellingham, WA: FNO Press, 1999.

McKenzie, Jamie. **Planning Good Change with Technology and Literacy**. Bellingham, WA: FNO Press, 2001.

Means, Barbara. "Critical Issue: Using Technology to Enhance Engaged Learning for At-Risk Students." North Central Regional Educational Laboratory. http://www.ncrel.org/sdrs/areas/issues/students/atrisk/at400.htm , 1997.

Moore, Geoffrey A. **Crossing the Chasm: Marketing and Selling High-Tech Products to Mainstream Customers**. New York: Harper Business, 1991.

Murphy, Carlene U.and Lick, Dale W. **Whole-Faculty Study Groups: A Powerful Way to Change Schools and Enhance Learning**. Newbury Park, CA: Corwin Press, 1998.

Senge, P. **Schools That Learn: A Fifth Discipline Fieldbook for Educators, Parents, and Everyone Who Cares About Education.** New York: Doubleday, 2000.

Slowinski, Joe. "Finding Your Way through The Data Smog." **From Now On**, September, 2000. http://fno.org/sept00/data.html

Technology Counts '99. Education Week, 1999. http://www.edweek.org/sreports/tc99/articles/summary.htm

Technology Counts 2001. Education Week, 2001. http://www.edweek.org/sreports/tc01/

Technology in Education 1999. Shelton, CT: Market Data Retrieval, 1999.

Trotter, Andrew. "Preparing Teachers For the Digital Age." **Technology Counts '99**. **Education Week**, 1999. September 23, 1999. http://www.edweek.org/sreports/tc99/articles/teach.htm

U.S. Department of Education. "Getting America's Students Ready for the 21st Century: Meeting the Technology Literacy Challenge." 1996. http://www.ed.gov/Technology/Plan/NatTechPlan/

Van Dam, Jan. "Total Cost of Ownership." **Technology and Learning**. October, 1991.

Wiggins, Grant and McTighe, Jay. **Understanding by Design.** Alexandria, VA: ASCD, 1998.

How is searching for the truth different from searching for nothing more than the facts?

In order to be truthful.
We must do more than speak the truth.
We must also hear truth.
We must also receive truth.
We must also act upon truth.
We must also search for truth.
The difficult truth.
Within us and around us.
We must devote ourselves to truth.
Otherwise we are dishonest
And our lives are mistaken.
God grant us the strength and the courage
To be truthful.
Amen

© 1991, Michael Leunig. From his collection, **The Prayer Tree** (ISBN 1-86371-034-5, Harper Collins Religious, republished here with permission from the author.

How do we discover truth in the world around us?
How do we teach students to discover their own truth or truths about the most important questions and decisions in life?

• What kind of city is this?
• What kind of country is this?
• Where should I live?
• What kind of person is this?
• What kind of novelist, poet, or essayist is this?
• What makes a great poem?

- Is this a great poem?
- What makes a great park?
- Is this a great park?
- What makes a great doctor?
- Is she or he a great doctor?
- What makes a good treatment decision?
- Is this treatment a good one for me?
- How can I make this city or country better?
- How can I improve upon this poem?
- Which poet, painter, sculptor, juggler, or lyricist would make the best mentor for me?

Answering any of these questions requires more than the mere collecting of facts. As Sven Birkerts wrote in his **Gutenberg Elegies**:

Resonance — there is no wisdom without it. Resonance is a natural phenomenon, the shadow of import alongside the body of fact, and it cannot flourish except in deep time.
(Birkerts - Page 75) **The Gutenberg Elegies** by Sven Birkerts ISBN 0-449-91009-1 Fawcett Columbine, NY, 1994

Schools often stop short of showing students how to fashion what Birkerts calls "import." The collection of facts without much purpose is trivial pursuit. (see **FNO**, February, 2001, "From Trivial Pursuit to Essential Questions." http://fno.org/feb01/covfeb.html)
We should be teaching students how to move past collection and gathering to insight, understanding and the discovery of import. We should be showing them how to discover "the heart of the matter."
Schooling should be a matter of discovering *lux et veritas* - Yale University's motto ("light and truth.") The motto *lux et veritas* is taken from Psalm 43:3 "Send out your light and your truth, let these be my guide."
Young people do not become wiser simply by amassing huge mountains of data, as we can see when we look at the challenge of studying any city.

Facts without Soul and Character

What kind of city is New York? is Sydney? When we read about a city in an encyclopedia, are we coming to understand the true New York? the true Seattle? the true Sydney? the true Melbourne or Adelaide? Try reading about a city that you know well (http://

dir.yahoo.com/Reference/Encyclopedia/).

Do you recognize the city you know in the article you read? What is missing? What kinds of information are left out?

If I asked you to supply three words that would best describe the character of a favorite city, what words would you use?

Do any of these words appear in the encyclopedia article?

I asked two friends what words they would apply to nearby Seattle. One friend replied, "Young, suburban and low-key." Another friend replied, "Salmon, Pike's Market and Pearl Jam."

When most people think of cities, they think of its qualities, traits, moods, special attractions, and styles. They think about the experience of living there or visiting there. They might want to know if they can find jazz with soul, top performers, and audiences that make the listening easy and pleasurable. They care less about the number of jazz clubs, art museums, or parks than the quality of these features.

Icons and More Icons

The Seattle Space Needle is an icon well known to many who have never visited the city. Do icons speak truthfully? But Seattle is a major port for shipping to and from countries along the Pacific Rim. Huge cranes lift cargo from the freighters that fill the port. How many people see or know this face of Seattle?

How about travel guides? Do they tell the true story? If we read about New York or Adelaide in a tourist booklet, will we uncover the real city?

Read about New York City at Yahoo - Yahoo Travel - http://travel.yahoo.com.

- Is New York friendly or cold to strangers?
- Is New York dangerous? safe? in between?
- Can you find good parks? museums? jazz? pizza? Samaritans?
- Does New York have soul? or must we walk down these mean streets without hope of redemption?
- Is Queens part of New York? Is Brooklyn?
- Does New York have a conscience?
- Would the homeless agree?
- How good are the ferries? the hot dogs? the baseball teams?
- Can you find gentleness? generosity? joy? a dog run?
- Is New York a good place to visit? to live? to avoid?
- Did Frank Sinatra really understand New York?

- Does New York have serious racial problems?
- Is New York a good place for children?
- How did 9/11 change New York? In what ways has it remained unbowed and unchanged despite the tragedy?

With the possible exception of the Frank Sinatra and 9/11 questions, most of the questions above could be asked about any city.. But the answers are rarely found in travel guides or encyclopedias.

- Is Sydney friendly or cold to strangers?
- Is Sydney dangerous? safe? in between?
- Can you find good parks? museums? jazz? pizza? Samaritans?
- Does Sydney have soul? or must we walk down these mean streets without hope of redemption?
- Is Manly part of Sydney? Is Cronulla?
- Does Sydney have a conscience?
- Would the homeless agree?
- How good are the ferries? the hotdogs? the football teams?
- Can you find gentleness? generosity? joy? a dog run?
- Is Sydney a good place to visit? to live? to avoid?
- Did Frank Sinatra really understand Sydney?
- Does Sydney have serious racial problems?
- Is Sydney a good place for children?

The Tourist Side of Town

Even if you visit a city like New York or Sydney physically, you may not come to know the "real city" very well, as most tourists taste only a small slice or sample of such cities. Midtown New York, with its hotels, theaters and restaurants catering to tourists, is a far cry from the Upper West Side with its multi-ethnic populations and intriguing brownstones. Cross to the East Side through Central Park and one's impressions change once more, as brownstones give way to soaring apartment buildings that seem populated by folks who stride briskly with brief cases, cell phones, and poodles.

In Sydney, many tourists spend their time down at Circular Quay in an historical neighborhood called the Rocks. They wander past ferries and street performers on their way to the famed Opera House, all the while enjoying a diet of visual icons they may have tasted long before arriving in Australia. With just a few days to visit, they are unlikely to stray far from the beaten path.

It is difficult to build an understanding of a city, but it is easy to collect facts. How many museums? How many acres of park? How many people live in the city? What is the homicide rate? What are the icons?

Secondhand Truth

But even though understanding almost any complex aspect of human experience is difficult, the importance of knowing how to fashion a deep and comprehensive understanding is critical. If we or our students cannot do this kind of thinking, we become prisoners to the secondhand smoke of propagandists, advertisers, promoters, and marketeers.

Without the skills to build and discover import, we may visit cities and simply end up taking the tour. We may spend our time in disneyfied neighborhoods that bear little resemblance to the original city. We see the city they want us to see. We buy a ticket and go for a ride. The city as theme park!

Without the skills to build and discover import, we let doctors, politicians, insurance agents, politicians, morticians, and pundits explain our world and our choices to us. We see the life they want us to see. We buy a ticket and go for a ride. Life as theme park!

If we do not learn how to dig deeply, consider wisely, and ponder the meaning of things, we are condemned to a kind of intellectual skating. We glide along the surface. We slide into mentalsoftness. We surrender. We give up the search for our own truths.

Prime Indicators of MentalSoftness™

1. Fondness for clichés and clichéd thinking - simple statements that are time worn, familiar and likely to carry surface appeal.
2. Reliance upon maxims - truisms, platitudes, banalities, and hackneyed sayings - to handle demanding, complex situations requiring deep thought and careful consideration.
3. Appetite for bromides - the quick fix, the easy answer, the sugar coated pill, the great escape, the short cut, the template, the cheat sheet.
4. Preference for platitudes, near truths, slogans, jingles, catch phrases, and buzzwords.
5. Vulnerability to propaganda, demagoguery and mass movements based on appeals to emotions, fears, and prejudice.
6. Impatience with thorough and dispassionate analysis.

7. Eagerness to join some crowd or other - wear, do and think what is fashionable, cool, hip, fab, or the opposite, or whatever . . .
8. Hunger for vivid and dramatic packaging.
9. Fascination with the story, the play, the drama, the show, the episode, and the epic rather than the idea, the question, the argument, the premise, the logic, or the substance. We're not talking good stories or story lines here. We're talking pulp fiction.
10. Fascination with cults, personalities, celebrities, chat, gossip, hype, speculation, buzz, and blather.

Mentalsoftness is a term coined by Jamie McKenzie in May, 2000. (See FNO, May, 2000, "Beyond Information Power."). It was inspired by hearing an Australian use the term "microsoftness."

Who cares?

Some times we ask students to do research that has little to do with verity. It may cast little light on the subject. It may not increase understanding or promote insight. It does not pass the test of "So what?"

Collections of facts might not tell us much.

How big are the parks? How many museums? How many people?

Students may spend two weeks gathering information about a city or a country without coming to know the spirit and the character of that city or country at all.

Knowing the number of museums is not the same as tasting them.

Reading the crime rate across the city is not the same as strolling down its streets, noticing its street corner society, its homeless population, its workers, and its families on the move.

Many of the other questions posed earlier in this article will take us beyond information to meaning . . .

Does Sydney offer good parks?

A good park? What does that mean? What are the traits or characteristics of a good park? How would I know a good park from this other side of the ocean?

Before we can judge the quality of New York's Central Park or Sydney's Botanical Gardens, we would have to define "goodness" when applied to parks. We ask our students to open a cluster-diagramming program like Inspiration™ and have them map out the traits of a good park . . .

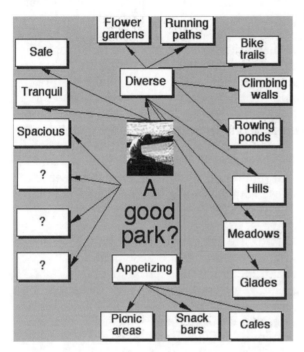

It turns out that knowing a city's character is a profoundly complicated task. Even those who live in a city may only know part of the city - their neighborhood and those aspects of the city they can appreciate through whatever filters they wear.

The factory worker may appreciate the fine selection of pubs while the banker may have missed that aspect entirely. A different factory worker may visit gardens rather than pubs. A different banker may be fond of watching football while pub crawling with her friends.

It turns out that just as "Beauty is in the eye of the beholder," determining the nature of a city is a personal judgment. One person hates Sydney but loves Melbourne. Another hates Melbourne but loves Sydney.

The reasons for these feelings rarely have to do with the facts available about cities in encyclopedias.

Virtual Social Studies

We must guard against launching virtual social studies - the apparent study of other lands and cultures that amounts to little more than a glib and superficial dance through basic facts, stereotypes, icons, curios, and misconceptions. Such studies may contribute little to the understanding of other lands - may actually distort the understanding of young people so that they come to see places like Australia as an extension of Outback Steak House or places like the United States as an extension of **NYPD Blue**.

During a recent visit to a school district, I glanced through a social studies text that purported to introduce 6th grade students to regions and countries around the world. Because I have been fortunate to visit Australia and New Zealand many times during the past three years, I was curious to see how this book might treat those countries.

The treatment was superficial, thin, and cursory. There simply wasn't enough space devoted to either country to do them justice. It was a gloss. The book shared some basic facts, made some geographical observations but ended up casting little light on either country.

But textbooks are not the only problem. Many of the Internet sources used to study foreign lands are similarly shaped by tourism and entertainment values that rarely allow for exploration of deeper cultural or social issues. These sources rarely look on the darker side of these cultures. Crime? Drugs? The mistreatment of women? of indigenous peoples?

Human rights and quality of life issues often fall into the shadows as we learn of great shopping, native foods, and beachside villas.

Social studies becomes a trek, a tour, an entertainment.

What kind of city is LA? Do we visit Hollywood? Disney? The beach?

Or do we look more carefully at the city and its issues? The sites of previous riots? The treatment of citizens by police?

Is life getting better in LA? for whom? since when?

What are people doing to make life better in LA? Which strategies have the most promise?

How will California's power crisis change the lives of various types of citizens? How will skyrocketing electric bills influence small business people? big business? wealthy families? low-income families?

What would you do if you were mayor of LA? police commissioner?

An Example of Learning for Understanding

In an excellent program called "Looking At Ourselves and Others," http://www.peacecorps.gov/wws/guides/looking/index.html the Peace Corps offers lesson plans that engage students in a deeper exploration of culture than the gloss mentioned above. The site provides resources for teachers and students that invite thinking with a global perspective about people and life in other countries.

Verity - the Search for Difficult Truth

Failing the Standards

Shallow, "entertainerized" learning about other countries or cities is indefensible. We create the illusion of learning, but the exploration is superficial and the outcomes untrustworthy. This approach does not sustain the kinds of analysis reflected in the following curriculum standards from Australian and American states:

1. Connecticut - "Describe the influence of U.S. political, economic and cultural ideas on other nations and the influence of other nations' ideas on the United States; analyze and evaluate the significance of major U. S. foreign policies and major international events and conditions over time; develop proposals regarding solutions to significant international, political, economic, demographic, or environmental issues."

2. Western Australia - "Students understand their cultural, geographic, and historical contexts and have the knowledge, skills, and values necessary for active participation in life in Australia. Students interact with people and cultures other than their own and are equipped to contribute to the global community. "

3. Illinois - "A study of social systems has two important aspects that help people understand their roles as individuals and members of society. The first aspect is culture consisting of the language, literature, arts and traditions of various groups of people. Students should understand common characteristics of different cultures and explain how cultural contributions shape societies over time. The second aspect is the interaction among individuals, groups and institutions. Students should know how and why groups and institutions are formed, what roles they play in society, and how individuals and groups interact with and influence institutions."

4. South Australia - INTERDEPENDENCE - "What knowledge, skills and dispositions are required to critically understand the systems to which lives are connected and to participate positively in shaping them? Learners develop:"
  • a sense of being connected with their worlds
  • capabilities to contribute to, critically reflect on, plan and take action to shape, local and global communities.
  This includes:
  • understanding cultural and global connections, patterns and evolutions
  • understanding what is needed for sustainable social and physical environments

• acting cooperatively to achieve agreed outcomes
• taking civic action to benefit community.

Understanding by Exploring Deeply

It is tempting to structure student learning so thoroughly that we leave little room for surprise, discovery and authentic idea building.

A curriculum design model currently fashionable in the States (**Understanding by Design** - Wiggins and McTighe, 1999) urges teachers to build curriculum units by identifying the essential questions and understandings envisioned by the curriculum standards in order to build backwards from these desired outcomes. If you know what you want students to know by the end of the unit - so the theory goes - then you can create a series of learning steps that will deliver the goods.

While this strategy is appealing on the surface, it might lead to a kind of teacher structuring and control that actually deprives students of more authentic learning.

In studying a city, for example, a teacher might have drawn five generalizations from the textbook:

1. New York is a big, high-energy city with lots of people and products.
2. New York was shaped by its location.
3. New York is an entertainment and banking center that shapes culture across the United States and the world.
4. Life in New York is getting better.
5. New York is a GREAT place to visit.

On the surface, these kinds of understandings may seem standards-based.

The following statements are drawn from Illinois state standards:

C. Understand relationships between geographic factors and society.

17.C.1a Identify ways people depend on and interact with the physical environment (e.g., farming, fishing, hydroelectric power).
17.C.1b Identify opportunities and constraints of the physical environment.
17.C.1c Explain the difference between renewable and nonrenewable resources.

17.C.2a Describe how natural events in the physical environment affect human activities.
17.C.2b Describe the relationships among location of resources, population distribution and economic activities (e.g., transportation, trade, communications).
17.C.2c Explain how human activity affects the environment.

But students must have a chance to wrestle with these kinds of issues in order to become thinkers. Some of the generalizations listed about New York above are highly questionable, but how likely is it that students will be given an opportunity to challenge or dispute the teacher's unit goals?

In her chapter, "Using Technology to Enhance Student Inquiry," in **Technology in Its Place**, Debi Abilock describes an approach to launching student investigations that is deep but practical:

> In problem based learning, students are confronted with ill-structured problems that mirror an authentic situation. It is important for them to recognize that just as in real-world problems, there are no simple right and wrong answers.
>
> (Page 4)

Abilock's chapter provides an excellent model for launching authentic discovery that addresses the skills listed in state standards while putting the primary responsibility on students to create meaning. Teachers will find during these kinds of investigations that most members of the class (teacher included) will emerge with surprising new understandings that could not (and should not?) have been foreseen and preplanned.

While structure and scaffolding are attractive (See "Scaffolding for Success," **FNO**, December, 1999), we must be on guard against imposing our preconceptions (or the textbook's bias) on our students. We must leave room for surprise and discovery.

Can we allow students to develop their own views of New York, Sydney, LA, and Melbourne? Can we trust them to discover, to build, and to generate fresh views of these cities?

References and Resources

Abilock, Debi. **Technology in Its Place**,
The Curly Flat - A web site devoted to Michael Leunig offering a brief biography, examples of his cartoons and merchandise. http://www.curlyflat.net/

Leunig Cards and Mugs - http://www.dynamoh.com.au/html/leunig_index.html

Wiggins, Grant and McTighe, Jay. **Understanding by Design.** Alexandria, VA: ASCD, 1998.

How can we best equip students to build their own meanings? To find their own truths? To pass beyond secondhand truths or cut-and-paste thinking?

We can combat the virtual social studies described in the previous chapter by equipping students with critical thinking skills that will empower them to ferret out the darker truths and strip off the pancake makeup that often conceals the reality of life in any city.

While the examples and the focus in this chapter may be upon social studies, the challenge of finding truth obviously crosses over into the other disciplines as well.

Mapping Out a Real City

If students begin their study by creating a complete mind map identifying all of the facets of a city, they are more apt to emerge from their investigations with a full picture. They are less apt to take at face value the portrait offered by the city and its promoters.

It might help if they learn about one city as a whole class first so they can appreciate how all of the facets combine to shape the identity of the city. Once they have studied New York or Sydney with the guidance of a skillful teacher, they may be able to apply a similar set of facets to Hong Kong or Portland.

If students start by gathering everything they find about a city, they might build a huge pile yet be completely unaware that big chunks and entire categories of information are missing. They only know that they have big piles and tend to equate the size of piles with the growth of understanding. These piles may tell them little about the true character of the city, but longstanding school rituals have placed little emphasis upon the gathering of small amounts of pertinent information.

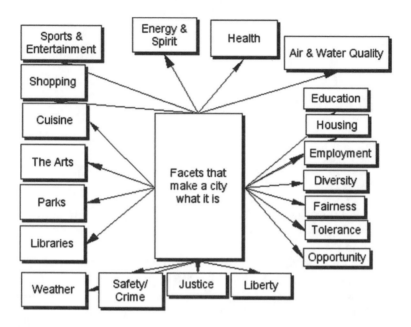

What's Missing?

If students do a good job of creating subsidiary questions to match each of the categories in the diagram above, the nature of research is changed radically.

Under "Weather," the student might ask a dozen questions such as:

1. What is the average annual rainfall?
2. How many sunny days annually?
3. How cold does it get during winter?
4. Are there frequent dangerous storms?

Under "Housing," the student might ask a dozen questions such as:

1. Are there enough housing units so that all citizens can live in healthy, safe units?
2. Has the city eliminated most "slum" housing?
3. Are housing costs reasonable?

4. Is there a big gap between low and high income housing?

With such a full listing of important questions, the search for understanding will focus upon filling in the blanks.

The Advantage of Juxtaposition

This kind of study works best when students compare and contrast several cities on the facets in the table. Without a comparison, it might be hard for the student to appreciate the import of the data being collected.

How many sunny days would be good?
How many homicides per hundred thousand represent a problem?
When can we say a city offers fair housing?
When does air pollution become a serious hazard?

Juxtaposition helps to provide a context within which to make judgments. Without these comparisons, the facts about crime, weather, and housing may not mean much to the students.

The Advantages of Teaming

Collecting sufficient data on all the key facets listed in Figure 17.1 is an enormous challenge - one well suited to team efforts. A teacher might ask four students to carve up the list so that each one gathers information about 4-5 facets about two cities such as Sydney and Perth. As each student becomes expert in certain facets, the team begins to share findings and decide their combined import.

The Veracity Model

We equip students with six or more questions to help them measure the level of veracity they have attained through their research efforts.

1. Did I make a list of qualities and issues the city fathers and mothers might want to keep in the shadows?
2. Did I deconstruct messages and ads to identify classic distortion strategies?
3. Did I locate critics, unusual sources, and critical comments?

4. Did I challenge claims by checking facts?
5. Did I look for evidence to the contrary?
6. Are my findings and conclusions anchored in facts?

Types of Bias

What are the main filters that might block students from a truthful view of a place? How can students learn to look beyond these filters?

Compare and contrast the filters listed below. Which is most apt to distort or block the understanding of your students?

1. Propaganda

Information is carefully selected to create impressions about the quality of life for citizens. This kind of biased treatment of information is also intended to sway decisions by appealing to the emotions (fears, dreams, etc.) Propaganda often uses partial truths, exaggeration, and related tactics to stir an audience. The source of propaganda is usually a political group, the government itself, or some interest group with an ax to grind.

2. Marketing

Marketing may use some of the same tactics as propaganda, but it is motivated usually by market interests, trying to sell products and experiences. Propaganda is more about selling ideas and action proposals. Marketing shapes information to satisfy the dreams, wishes, and needs of the consumer. In an effort to win tourist confidence, for example, a city might play down any recent problems with street crime.

3. Class

How we see or report the world may differ according to our place in that society. A judge or banker might live high on a hill and see life from up in the clouds while a street sweeper might have a different perspective, different air quality and different water quality.

If the reporter is content with life and her/his position within the society, the resulting stories and information may be slanted by those attitudes and life experiences. Insulated from the suffering and difficulties of life at the bottom of the social pyramid, the reporter may describe the city through "rose-tinted glasses."

Conversely, someone with much life experience at the bottom of the pyramid may devote all of their attention to the slime, pollution and social problems that bombard them day after day. Roses? It may have been a long time since they have had a chance to smell or see them.

4. Ethnocentrism

There is always some danger that we will read or translate information through the blinders, lenses, and attitudes of our own culture. Our students may be locked into the values of their own town or group or nation. They may be a bit swift to judge other groups or cultures from a superior vantage point. They may not view the new and different culture as a potential source of inspiration.

Related to this might be the creation of information about a foreign city or country by our own country fellows - an Australian view of San Antonio, a Canadian view of New York, an American view of Hong Kong. The source may not employ local people to create the information at all.

Smart chefs allow for fusion. They allow the influence of other cultures to enrich their own.

5. Ignorance

Much less intentional than the previous types of bias would be information flawed by the creator's lack of background knowledge, training, expertise, and judgment. Some people do not know what they do not know. They pose as an authority without owning any of the qualities that should accompany that status. Anyone can create a Web site and pretend to know something.

Can we encourage our students to challenge the authority and value of the information resources they encounter, whether they be print, analog, digital, or spoken?

Limitations of Mainstream Sources

When students first turn to the Net to learn about cities and nations, what kinds of information, images, and conceptions predominate?

Take a look at what types of information the following four sources have to offer about Hong Kong (or any other city) and note the limitations of this type of information.

Other Worldly Learning

What are the strengths? the weaknesses?

• Source One: Yahoo
http://search.yahoo.com/

• Source Two: KidsClick
http://sunsite.berkeley.edu/KidsClick!/

• Source Three: Google
http://www.google.com

• Source Four: CIA Data
http://www.cia.gov/cia/publications/factbook/

The reader will quickly see that these sources tend to emphasize the tourist's and westerners' view of foreign cities such as Hong Kong. Much Web development is funded with marketing money aimed at attracting visitors. As a result, much of the information offered has been sanitized or eliminated. It may be difficult to find information about bad water, hygiene or air quality from such sources.

Ethnocentrism is the tendency to view and judge other cultures through the values, perspectives and bias of one's own culture. In the worst of cases, this approach can damage relations between the two cultures, as when a business person or diplomat fails to study the culture and traditions of a host nation with proper respect and badly mangles exchanges by showing an ignorance of proper rituals and behaviors. In other respects it means blocking one's mind to the many deep lessons to be learned by exploring different cultures with an open and appreciative mind. Many Western business leaders, for example, have grown to embrace and employ some of the tactical approaches associated with Taoism, adapting the best of Eastern culture to enrich business life at home.

In addition to the marketing focus and ethnocentrism of many conventional Web sources, some of the so-called child-safe sites such as KidsClick and Yahooligans rely on Web sites created by American school children to tell us about other countries such as Australia. While it might be fun for students to read what other students think about Australia, these sources hardly meet the test of authority and reliability. We would hardly want to build our research on foreign cities upon the writings of young people who have never actually visited those cities.

Other Worldly Learning

Alternative Sources - Triangulation

What are the best ways to equip students with the predisposition and the skills to challenge conventional wisdom and marketing claims - an informed but constructive skepticism? How can multiple sources protect against propaganda and distortion?

Have we been looking for truth in all the wrong places?

Where should we be looking? Where do we want our students to look? No matter what the city, the topic, or the question, there are strategies likely to help researchers find alternative sources and points of view.

1. Ask the critics.

Take the time to figure out who might not agree with the party line, the conventional wisdom, and the way things are "spozed to be."

Sadly, local newspapers can generally be counted upon to serve up generous helpings of slander and trouble, but their articles are often missed by search engines. Students might proceed directly to a global listing of newspapers such as http://www.totalnews.com/.

While the critics may not be prominently listed on major Internet listings, they can often be found my combining a major search term such as "Hong Kong" with critical verbs or nouns such as "dispute, controversy, failings, problems, issues, etc."

We can teach students to look for Web sites, listservs, publications, and other sources where critics may vent their feelings.

2. Ask the helpers.

When it comes to the darker side of life, helping agencies are often inclined to paint life in vivid terms, in part because this strategy often leads to bigger donations and more commitment to the cause at hand.

Just how bad is the homeless problem? Ask the Salvation Army.

3. Contact real people real time.

E-mail makes it possible to ask questions directly to private citizens, experts, government officials, school children, and senior citizens.

Chapter 19 - Going Unplugged

Knowing when to go unplugged is certainly one of the preeminent skills of this new century - knowing when to go natural and when to turn to print and hard copy materials instead of digital offerings.

Some of us sail, drive or hike away on summer holidays designed to spare us from the pressures of e-mail and other digital influences. We turn off cell phones and listen to gulls calling to each other.

But new sailboats come equipped with GPS and computer screens that sit in front of the ship's wheel. We may find ourselves glued to the screen watching digital islands moving on a chart as the real islands glide by unseen on port and starboard.

Can we keep the technologies where they belong?

Schools may already suffer from too little intimacy, so we should be cautious about pressures to embrace distance learning, virtual learning, or virtual schools. There are many times when digital comes up short despite the lavish claims and promises of companies who wish to sell us digital adventures, rain forests, lessons, and meetings.

Afraid of flying since 9/11? Technology companies quickly suggested video-conferencing as a substitute for face-to-face. But chemistry, eye contact, and proximity all play major roles in effective bargaining, problem-solving, and negotiating.

Reach out and touch someone? Digitally?

As the next section illustrates by reviewing the treatment of important thinkers by the Net, digital is not always best.

The HotBot Top 40 Revisited

In 1996, just as the Internet was becoming a major new source of information for schools and the general public, I conducted a review of the Net coverage afforded to major figures. Using HotBot as a search engine, I looked at and reported the number of Web pages devoted to each of 300 people or religious figures.

Does the Internet provide solid information about the great thinkers and leaders of our times? Who gets the most attention? Heroes? Sports figures? Entertainers? Celebrities? Leaders? Thinkers?

Back in 1996, celebrities won the major share of Web attention. Thinkers and important leaders took a back row seat to film stars, singers, sports figures, and those with recent scandals on their list of credits.

In October of 1999, the same trend was reported as Madonna, Elvis, Bill Gates, and Michael Jordan scored right near the top along with Buddha, Kennedy, and Lincoln. Both Frank Sinatra and Princess Diana, who had passed away shortly before the list was compiled, rose dramatically in the rankings. Important thinkers such as Rollo May, Julian Huxley, Indira Ghandi and L. S. Vygotsky were virtually ignored.

In October of 2001, the Net continued to lavish attention on celebrities and headline grabbers. The full listing of results can be found at http://fno.org/Oct01/unplugged.html#hotbotlist.

Madonna has jumped ahead of Jesus to the Number Two spot. George W. Bush and Pamela Anderson each improved their rankings by more than 50 slots. Much of the list is unchanged.

Those who work quietly and seriously out of the limelight creating the best ideas and the intellectual capital for this knowledge-based society are rarely recognized. As a library substitute, the Net is sadly lacking. Its collection development is often driven by ratings and popularity.

How did Edna St. Vincent Millay Really Die?

Especially in categories such as biography, the most reliable and carefully researched information often still resides in printed books such as Nancy Milford's biography of Edna St. Vincent Millay, **Savage Beauty** (Random House, 2001).

Ms. Milford began her research on Millay in 1972. Twenty-nine years later, after carefully combing through original documents and letters and after interviewing dozens of Millay contemporaries, her research emerged as a printed biography. This work was not something to be posted and given away freely on the Internet.

If you want to know about Millay, you will want to buy this book or borrow it from your library. It works the same way with most

Going Unplugged

serious biographies of important people.

The Net casts little light on Millay's life. The pickings are slim - 16,500 Web pages - some well intended sites, but little illumination or professional scholarship. In many cases, the authors of these sites never reveal their identities, their background, or their education. The Net is generous in serving up her poetry, but biographical sketches are generally thin, usually brief, and often amateurish. Worse, they are frequently mistaken as to some basic facts such as the circumstances (and spelling) of Millay's death.

One learns several versions of her death . . .

Millay, a smoker in an age of smokers, succumbed to hear failure in 1950 at her home, Steepletop, in Austerlitz New York.

http://members.aol.com/MillayGirl/bio.htm

After her husbands death, she went on living in their isolated house in Austerlitz and died* there alone of a heart attack in 1950.

*Millay died of a broken neck suffered in a suspicious fall down the stairs at Steepletop. As a founding member of the Millay Society, I have had the pleasure to not only meet with Norma Millay (her sister, who passed on in 1986) but with her biographer Nancy Milford, whose book on Millay is 20+ years in the making and probably being withheld pending the death of a possible accomplice to her suspicious death. Nancy actually interviewed the doctor who wrote Millay's death certificate after she found that it said, Suspicious fall down stairs, broken neck. I have performed a one-woman show on Millay all over the country and my master's thesis was about her relationship with Edmund Wilson, so I am well-versed in her life and work. Thanks for a fun bio on her. Hope this helps. All the best, JKH

http://www.geocities.com/Paris/LeftBank/6865/millaybio.html

Nancy Milford's printed, full-length biography does confirm that Millay died by falling down the stairs and quotes the doctor's note.

151

Going Unplugged

But the majority of the biographical sketches available on the Net pass along the heart attack story as if it were factual.

Millay, who with her husband had drunk to excess since the 1930s, evidently grew more dependent on alcohol during her brief, inconsolable widowhood. She died sitting at the foot of her staircase, alone, at Steepletop.
http://www.english.uiuc.edu/maps/poets/m_r/millay/millay_life.htm

They were together until his death in 1949. Millay died the following year of heart failure. http://www.theglassceiling.com/biographies/bio44.htm

---

The Web's coverage of the best thinkers and artists from the previous century leaves much to be desired. For the most part, the important, thoughtful, serious scholarship about these people still resides in libraries and printed books. Much of their work and thinking is found in the same location.

The New Economy, The Virtual Library, and Other Myths

The **New York Times** has reported that the corporate push to replace printed books with electronic books has stalled and fallen well short of predictions.

Forecasts of an E-Book Era Were, It Seems, Premature
By David D. Kirkpatrick
(from the August 28, 2001 **New York Times**,)

Laurence Kirshbaum, chairman of the books division of AOL Time Warner (news/quote), pledged to lead the charge:

"We want to see electronic publishing blow the covers off of books. Andersen Consulting had recently estimated that by 2005 digital books could account for 10 percent of all book sales."

A year later, however, the main advantage of electronic books appears to be that they gather no dust. Almost no one is buying. Publishers and online bookstores say only the very few bestselling electronic editions have sold more than a thousand copies, and most sell far fewer. Only a handful have generated enough revenue to cover the few hundred dollars it costs to convert their texts to digital formats.

Along with the demise of the so-called New Economy, the proponents of which spoke sneeringly of the Old Economy, it may be timely to reconsider the bombastic predictions of electronic libraries and the Net replacing the need for school and public libraries.

Some who pose as futurists are little more than salespeople and cheerleaders for a digital product line that has severe limitations, shortcomings and weaknesses. The attempt to create peer pressure for this product line surfaces as lifestyle advertising equating digital connectivity with freedom and joy, yet evidence accumulates that not all things digital are healthy, reliable, or worth buying.

These days it is fashionable for some to push 21st. century skills on schools that seem tilted toward digital living and high levels of technology penetration, but many of these messages are based on unexamined assumptions about the impact of these tools and experiences upon the quality of life and the quality of classrooms.

The skill of unplugging digital tools should rank highly on the list of 21st century skills when other tools and experiences prove superior.

There are times to turn off the equipment, shut down the covers of the laptops and focus on the dialogue and exchange of ideas.

We should help students learn to turn off their cell phones, their DVD drives, and their handheld personal assistants when they might interfere with a walk along a river, a dinner with a friend or the chance to curl up with a good book about Edna St. Vincent Millay.

Discerning use of new technologies and classical technologies is a skill whose time has come.

Deep Thinking, Pondering and Resonance

Resonance
-- there is no wisdom without it.
Resonance is a natural phenomenon, the shadow of import alongside the body of fact, and it cannot flourish except in deep time.

Sven Birkerts, **The Gutenberg Elegies**, Page 75

Some technologies do a better job of inspiring resonance and deep thinking than others. We do know that the hurried (sometimes fren-

Going Unplugged

zied) pace of modern life can interfere with the kind of reflective thought and pondering that leads to profound understandings and elegant inventions.

Long ago, someone created porches to encourage deep thought.
Someone else came up with park benches.
Another inventor saw the value of hot tubs.
Running shoes and walking shoes are often effective.
Even a treadmill can inspire deep thinking unless one dons headphones and watches the TV news while running.

Plugged?
Unplugged?

It is difficult to touch or find the deeper truths when the noise and chatter of modern life penetrates, interrupts, and distracts.
It is difficult to find our own voices, our own visions and our own possibilities while the cultural loudspeakers blast out shrill messages wherever we go.

Plugged?
Connected?
Distracted?

One commentator tells a tale of hiking deep into the mountains to find solace near a lake far from the traffic and noise of the city. He leaves his cell phone behind - unplugged and ready for an afternoon of deep thought. After a few minutes, just as he begins to enter a mood and thought space called "reverie" by some, a cell phone rings insistently across the lake as another hiker (still wired and connected) reaches out and touches someone with a voice that echoes across the lake and shatters the reverie.

Can we teach young ones to cherish park benches, long walks through the forest, and a day without e-mail or cell phones?

We must.

Chapter 20 - Beyond Bamboozelement

We are emerging from a troubling decade of bamboozlement and skullduggery. The technology and dotcom bubbles have burst, the fanciful forecasts have collapsed and many of the false prophets have been caught in the harsh light of publicity.

Cooking books and misstating earnings in order to line personal pockets has thankfully been stopped cold. But meanwhile, millions of people are looking at severely depleted pension funds and the prospect of delayed retirements. Thousands of others find themselves out on the streets as telecommunications companies stumble, and sales of technologies dwindle and stall.

As this book goes to press, the American stock markets remain depressed and volatile while the trust of investors is so uncertain that billions have been diverted to bonds and other investments. New financial scandals seem to pop up from week to week so that corporate earning reports are viewed with suspicion.

Just what is the true price/earnings ratio? This is a difficult question to answer when earnings reports may conceal billions of expenses and losses in order to put a best face forward.

Educators might ask some of the same questions about the benefits of new technologies.

1. What is the real return on investment when a district connects all classrooms to the Internet at the cost of millions of dollars but spends little money on program and professional development?

2. How much of a day will teachers and students be able to use newly purchased laptops when the batteries last just half as long as promised?

3. How many years will classroom computers last before software companies introduce combinations of improvements and upgrades that render those computers obsolete and require premature replacement?

Beyond Bamboozelement

Dotcom Lessons

We are emerging from a period of frantic speculation. Many dotcoms have bombed. After several years of disdain for old time values such as profit, value and return on investment, many of the upstarts and renegades have depleted their cash reserves, folded their tents and hit the streets looking for work with the same companies they recently derided as "old economy" dinosaurs. Instead of early retirements with fat stock options converted into fortunes, many of the evangelists and prophets of the "new economy" have launched a search for pay checks and security.

There is a lesson here for educators. Many of the promises, projections and products being heralded by enterprising education dotcoms are truly exciting and worth a careful look. But others are ill conceived, violate essential educational philosophies and carry with them substantial risks.

Internet Hype and Bombast

Despite the claims, not all digital experiences are automatically beneficial or preferable to print experiences. One author and speaker popular on the technology circuit, makes the following statement in one of his books:

> Time spent on the Net is not passive time, it's active time. It's reading time. It's investigation time. It's skill development and problem-solving time. It's time analyzing, evaluating. It's composing your thoughts time. It's writing time.
> **Growing Up Digital** (Tapscott, 1999)

Those who work in schools and have watched students making use of the Net might quarrel with Tapscott's assertion, as many students will squander time on the Net doing many things not listed by Tapscott like chatting, downloading songs, and playing Doom.

It turns out that the Net may prove valuable and worthwhile in particular cases, but value is not an inevitable byproduct of mere exposure. Schools and classrooms do not improve simply because Internet connected computers sit in corners or because students surf the Net. Students do not automatically analyze, evaluate, and consider what they are browsing. They may snack. They may gorge. But thoughtful, discerning, and deliberate use is not a given.

Beyond Bamboozelement

Digital learning is not automatically better than other kinds of learning. Results depend upon the way activities are planned and structured. But mere exposure to the Net rarely works the miracles claimed by technology cheerleaders. To the contrary, we witness technology-related disturbing new phenomena such as "the new plagiarism" and tendencies toward powerpointlessness, mentalsoftness and glib thinking. Along with the Net comes the disneyfication of much information and a focus on edutainment and infotainment.

Making Wise Investments

Before committing huge sums to new enterprises, schools need to consider the likelihood of winning a major return on the investment.

Those leading schools must protect them from edutainment and infotainment. The most promising strategy is to apply a filter to any newly proposed venture.

Before adopting an innovation, the planning group should ask the following questions:

1) What evidence exists indicating that this venture might enhance student performance on demanding new state tests and prepare them to show mastery of new curriculum standards?

2) If there is no reliable performance data available to assess the value of this venture, what aspects might contribute to producing the following outcomes?
- Improved reading, writing, and thinking
- Collaborative problem-solving
- Persuasive communication
- Enriched world view

3) To what extent is the new venture likely to win broad acceptance by the teaching staff as a valuable element to be blended into normal programming? Will it seem practical, relevant and user friendly or will it appear peripheral, frustrating, and off task?

4) To what extent will the adoption of this innovation require a substantial shift in staff attitudes, skills, and behavior such that major investments will be required in the form of professional development opportunities?

5) Could the learning activities and outcomes involved in this venture be accomplished just as well (or better) using books and other technologies? Is there a case of technology for technology's sake?

Beyond Bamboozelement

A Focus on Value

An emphasis upon student learning and information literacy is probably the best protection against education dot bombs, bamboozlement, and skullduggery..

When schools ask how they might best teach students to analyze, interpret, and infer with these new tools, they dramatically increase the likelihood of realizing a real return on their investment dollars.

Online resources do not always provide the quality of good books and other print resources.

When we engage students as infotectives, expecting from them the same kind of thinking that Sherlock Holmes or Nancy Drew would employ to solve a mystery, we stand a good chance of improving performance on state tests.

- Finding meaning
- Creating meaning
- Extending meaning
- Reading between the lines
- Working with clues
- Building theories

In contrast, when we engage them in trivial pursuit or investigations that involve more entertainment than rigor and substance, we waste their time and risk inspiring the (warranted) resistance of staff members who already have too little time to address the demanding curriculum standards of this decade.

Traits of Effective Learning Products

School leaders may protect students and staff from educational bamboozlement by clarifying criteria and values. To qualify for investment, an educational program should meet traits like the following:

1. Challenging
Students are engaged in exploring essential questions requiring analysis, synthesis, and evaluation.

2. Standards-based
Activities are explicitly designed to develop the skills, understand-

ings, and knowledge required by state standards. The product offers direct links with standards.

3. Motivating

The product is designed in a way that will capture students' interest by sparking their curiosity and emphasizing essential questions, themes, and issues rather than relying upon glitzy entertainment strategies that emphasize an arcade approach to learning.

4. Attractive and User-Friendly

The digital pages and activities are designed to appeal to young people and support efficient navigation without surrendering to pop culture values.

5. Scholarly and Reliable

The information, the writing, and the learning activities are based on the work of professionals in the field of learning who have a deep and solid background in the material combined with an understanding of how young people learn, explore, and build meanings. The company supplies background of product developers.

6. Research-Based

The product is based on independent and reliable research findings, "best practices," and field tests of prototypes.

7. Enhanced

The product offers students new experiences, opportunities and types of learning that would not be possible with older technologies. These opportunities are not only novel but also valuable as preparation for ways of living and working that will be essential aspects of the students' futures.

8. Cost Effective

The product delivers opportunities at a reasonable cost - cost that stands in proportion to the value being delivered.

9. Sustaining and Richly Varying

The product offers many choices, varies learning experiences and appeals to different learning styles so that it can maintain a high level of interest over time. The product will not suffer from an early fading due to :gadget effects." The product is organized so as to support and encourage the development of independent learning skills and attitudes.

10. Assessable

The product provides a substantial battery of assessment tools and opportunities so that students may evaluate and then adjust the quality of their own learning.

Schools can measure the effectiveness of technology and literacy related programs by tracking evidence of change in the daily practice of participating teachers as well as the comments of students. Surveys like those in Appendix A and B help to develop evidence of growth, learning, and change.

Desired Behaviors

1. Daily appropriate use

As teachers gain in skill, confidence, and inclination, they and their students will begin reporting daily use of new technologies alongside more traditional technologies such as books, Post-It Notes™ and Magic Markers™.

2. Continued self-sustained growth

Participating teachers report sustained acquisition of new technology skills along with an expansion of lesson design capabilities.

3. Access for all students

Without exception, students in the classes of participating teachers report sufficient access to tools to manage assignments effectively.

4. Self-sustaining community of learners

Participating teachers indicate that adult learning is collaborative, ongoing and informal.

5. Technology invisible, transparent, natural

Artificial, silly uses of new technologies subside and technology for the sake of technology is no longer evident. Uses of new technologies are comfortable, casual, and unexceptional.

Signs of Effectiveness

6. Expanding definitions
Classroom strategies and activities move from an emphasis on technologies and technology skills to focus on information literacy, research, questioning and standards-based learning. Students are challenged to analyze, infer, interpret, and synthesize with a mixture of classical and digital tools.

7. Support for engaged learning
Students report that they spend an increasing percentage of their time on Engaged Learning tasks - taking responsibility for their learning in a collaborative, strategic, and energized mode.

8. Teacher as facilitator
Both teachers and students report that teachers devote an increasing proportion of their class time to facilitating and guiding the learning of students as opposed to more teacher centered activities.

9. Discerning use
Participating teachers gain enough in confidence and discernment to identify technology uses and activities they have discarded or found unsatisfactory. They report movement toward quality and worth.

Busted flat in Baton Rouge
Headin for the train
Feeling nearly faded as my jeans
Bobby thumbed a diesel down
Just before it rained
Took us all the way to New Orleans

from **Me and Bobby McGee** (Foster/Kristofferson)

Many schools that have attempted **Just in Time Technology** have found themselves like the characters in the song above. Battery life is a HUGE unstated issue and problem, as many batteries seem to last half as long as promised and take twice as long to recharge. Compounding these issues are battery memory management challenges far too complicated to survive amidst the stress and pressures of schools.

For a detailed description of battery memory challenges, read the excellent review, "Maintaining and Troubleshooting Your Laptop Battery" by Bill Platt at **Hardware Hell**:

Battery memory is where the battery becomes conditioned to run for less time than it is designed to run. Say, for example, you run your computer on battery for an hour and then you plug it back in to let it recharge. The battery will become conditioned to run only an hour before it runs out of juice.

To correct Battery Memory problems, you must completely drain the battery and recharge it.

http://hardwarehell.com/battery.htm

The majority of schools interviewed for this book had experienced severe battery problems. The short battery life nearly crippled program effectiveness in many places, as teachers responded to the problems by avoiding use of the laptop carts altogether.

Busted Flat

Half a Life is No Life at All

Picture a middle school teacher with five seventh grade social studies classes all doing world cultures in 40 minute classes. She wants her students to make use of Inspiration™ to organize research into the quality of life in foreign cities using some of the strategies outlined in Chapter Seventeen - "Other Worldly Learning."

If this teacher brings the laptop cart and its 15 computers into her classroom for the week, she needs the laptops to function almost all day long with very few interruptions. There may be a few brief segments between classes and during lunch when recharging of batteries can take place, but some laptops do not charge well unless completely shut down (a process that can take 6-8 minutes between shutting down and restarting).

Most of the schools interviewed stated that they were lucky to get 90 minutes from their batteries. In many cases they had witnessed isolated batteries dying much earlier than expected, frustrating both teachers and students immensely.

But even if the teacher could count on 90 minutes, how does she carry out her unit plan when those first 90 minutes are done? She may have reached two of her five classes by now, but she will have to put all the laptops back in the cart to recharge for 90-120 minutes while her third and fourth period classes do a non laptop activity.

If she is lucky, the batteries will be ready to go by her last class of the day.

How does she start the next day?

In order to give her third and fourth period classes a chance at the laptops, she leaves them unused during the first and second periods and then finds them drained by the end of the fourth period so that her last class of the day may not use them either.

Without some kind of battery system that lasts all day long, a laptop cart is really only half a laptop cart.

Without a battery solution, schools are realizing a poor return on their investment.

Seriously Penny Wise and Pound Foolish

How does this happen?

1. The school buys a cheap battery and laptop combination. As Bill Platt explains in the article mentioned earlier, NiMH batteries usually last 1.5 - 2.5 hours and LiION batteries usually last 2.0 - 3.0 hours. Ever mindful of costs, schools often opt for short battery life,

saving less than $100 per laptop and curtailing laptop use by 25-40 per cent.

2. The school fails to buy a second battery and charger to solve the problem. An external battery charger costs less than $50. A second good LiION battery costs between $100 and $150. By spending this extra $200 - $250, the school can run its laptops all day long and the teacher mentioned earlier can use her lesson with all five classes.

Ignoring the battery issue cuts the value of a $1,500 laptop in half, but the damage is more severe than mere loss of time and use. According to many of the project leaders interviewed, a large number of teachers simply turn their backs on the laptop carts because the battery issues and scheduling issues are viewed as overwhelming.

What keeps schools from spending the extra $200 - $250? Evidently, the extra expense is seen as extravagant. Crippling the laptop cart and losing half its daily use might seem like a less costly approach on the surface, but the true cost is $750 of lost laptop productivity and access along with the loss of program credibility and teacher buy in.

Wiring the Wireless Classroom

For some reason, when batteries become a serious issue, some schools try to solve the problem by bringing power strips into the classroom so that students may plug their laptops into power sources.

While this plugging into the wall might seem like a solution, it represents a serious retreat and compromise of the wireless laptop promise.

Because few classrooms offer multiple power points, plugging into the wall may severely restrict movement, tie students to particular sections of the room and eliminate many of the advantages associated with wireless, mobile computing. (see Chapter One.)

- Ease of movement
- Relaxed fit
- Strategic deployment
- Flexibility
- Cleanliness
- Low Profile
- Convenience
- Simplicity
- Speed

References

Becker, Henry. **Internet Use by Teachers**. (1999) http://www.crito.uci.edu/TLC/FINDINGS/internet-use/startpage.htm

Bloom, B. (1954). **Taxonomy of Educational Objectives. Handbook I: Cognitive Domain**. New York: Longmans, Green & Co.

Brown, J. and Duguid, P. (2000) **The Social Life of Information.** Cambridge: Harvard Business School Press.

Cuban, L. (1986) **Teachers And Machines: The Classroom Use of Technology Since 1920**. New York: Teachers College Press.

Cuban, L. and Kirkpatrick, H. "Computers Make Kids Smarter— Right?" **TECHNOS Quarterly For Education and Technology**, Vol. 7, No. 2, Summer, 1998. http://www.technos.net/journal/volume7/2cuban.htm

Cuban, L. (2002) **Oversold and Underused - Computers in the Classroom.** Cambridge, Harvard University Press, 2002.

Eberle, Bob. (1997) **SCAMPER**. Prufrock Press.

Fullan, Michael G. (1991) **The New Meaning of Educational Change**. New York: Teachers College Press.

Fullan, Michael G. (1996) **What's Worth Fighting for in your School.** New York: Teachers College Press.

Joyce, Bruce R. and Weil, Marsha. (1996) **Models of Teaching.** Needham Heights, MA: Allyn & Bacon.

Joyce, B. (Ed). (1990) **Changing School Culture through Staff Development**. Alexandria, VA: ASCD.

Lieberman, Ann and Miller, Lynne. (1999) **Teachers—Transforming Their World and Their Work**. New York: Teachers College Press.

Lieberman, A. (1995) **The Work of Restructuring Schools: Building from the Ground Up**. New York: Teachers College Press

Loertscher, David. **The Organized Investigator** - (Circular Model)

McKenzie, Jamie. (2000) **Beyond Technology: Questioning, Research and the Information Literate School**. FNO Press. http://fno.org/beyondtech.html

McKenzie, Jamie. (1999) **How Teachers Learn Technology Best**. Bellingham, WA: FNO Press. http://fnopress.com

Senge, P. (2000) **Schools That Learn: A Fifth Discipline Fieldbook for Educators, Parents, and Everyone Who Cares About Education.** New York: Doubleday.

Tapscott, Don. (1998) **Growing Up Digital.** New York: McGraw-Hill.

Technology Counts '99. Education Week, 1999. http://www.edweek.org/sreports/tc99/articles/summary.htm

References

Trotter, Andrew. "Preparing Teachers For the Digital Age." **Technology Counts '99**. **Education Week**, 1999. September 23, 1999. http://www.edweek.org/sreports/tc99/articles/teach.htm

Van Dam, Jan. "Total Cost of Ownership." **Technology and Learning**. October, 1999.

Appendix A - The Student EDP

1. I make use of networked computers to explore important questions and issues arising out of the content of this class.

never __ monthly__ weekly __ 2-3 times weekly __ daily __

2. My teacher challenges me to do my own thinking, build my own answers and interpret information, using new technologies only when they might prove helpful.

never __ monthly__ weekly __ 2-3 times weekly __ daily __

3. The Internet and e-mail makes it possible for me to communicate with experts, other students and people from around the world to enrich my learning.

never __ monthly__ weekly __ 2-3 times weekly __ daily __

4. I organize my thinking using Inspiration™ and other software programs to make mind maps.

never __ monthly__ weekly __ 2-3 times weekly __ daily __

5. I make smart choices about the tools I use to accomplish tasks, using books, a spreadsheet or digital information when each one is the best

strongly agree __ agree__ disagree__ strongly disagree __

6. I work in a group to solve problems, make decisions and explore challenging questions.

never __ monthly__ weekly __ 2-3 times weekly __ daily __

7. I have the software skills I need to handle classroom assignments and challenges effectively and efficiently.

strongly agree __ agree__ disagree__ strongly disagree __

8. I am getting quite good at recognizing worthy uses of new technologies while avoiding the silly, trendy uses that waste time without delivering much of value.

strongly agree __ agree__ disagree__ strongly disagree __

9. My teacher presents the class with a challenge or issue, points us to a large collection of relevant information resources and expects us to figure things out.

never __ monthly__ weekly __ 2-3 times weekly __ daily __

10. I am able to make an important contribution to the work of a team considering a curriculum challenge.

strongly agree __ agree__ disagree__ strongly disagree __

Appendix A

11. The work we do in class and the tasks we must perform are going to prepare me for my life as an adult - both as a worker and as a community member.

__ strongly agree __ agree__ disagree__ strongly disagree

12. Many of the things we do with technologies in our class seem kind of flashy and senseless to me.

__ strongly agree __ agree__ disagree__ strongly disagree

13. We have as much access to technology as we need and we can always get more if we need it.

__ strongly agree __ agree__ disagree__ strongly disagree

14. We don't bother to do much with networked information because the network is either too slow or is breaking down right in the middle of a lesson.

__ strongly agree __ agree__ disagree__ strongly disagree

15. Most days we do not need to make use of new technologies in my classes. They have no real place in this type of course.

strongly agree __ agree__ disagree__ strongly disagree

16. If I run into in a new task requiring software or technology skills I do not already possess, I am quite good at teaching myself the new skills or finding someone to help me learn them.

strongly agree __ agree__ disagree__ strongly disagree __

Appendix B - The Teacher EDP

1. I ask students to use networked computers to explore important questions and issues arising out of the content of my class.

never __ monthly __ weekly __ 2-3 times weekly __ daily __

2. I challenge students to do their own thinking, build their own answers and interpret information.

never __ monthly __ weekly __ 2-3 times weekly __ daily __

3. I encourage students to use the Internet and e-mail to communicate with experts, other students and people from around the world to enrich their learning.

never __ monthly __ weekly __ 2-3 times weekly __ daily __

4. I expect students to organize their thinking using Inspiration™ and other software programs to make mind maps.

never __ monthly __ weekly __ 2-3 times weekly __ daily __

5. I encourage and model smart choices about the tools students might use to accomplish tasks, using books, a spreadsheet or digital information when each one is the best.

never __ monthly __ weekly __ 2-3 times weekly __ daily __

6. I ask students to work in groups to solve problems, make decisions and explore challenging questions.

never __ monthly __ weekly __ 2-3 times weekly __ daily __

7. I have become quite good at adding to my technology skills by asking for help, using the tutorials and teaching myself.

strongly agree __ agree __ disagree __ strongly disagree __

8. I really don't have the time or energy to do much lesson or unit development for new technologies.

strongly agree __ agree __ disagree __ strongly disagree __

9. I take pleasure in learning new approaches alongside of my peers in ways that are informal, casual and low-key.

strongly agree __ agree __ disagree __ strongly disagree __

10. I rely mainly on packaged programs to make sure my students have good technology experiences.

strongly agree __ agree__ disagree__ strongly disagree __

11. There is so much curriculum content to cover that I can rarely take the time to engage students in group investigations and problem solving.

strongly agree __ agree__ disagree__ strongly disagree __

12. I am getting quite good at recognizing worthy uses of new technologies while avoiding the silly, trendy uses that waste time without delivering much of value.

strongly agree __ agree__ disagree__ strongly disagree __

13. If I run into in a new task requiring software or technology skills I do not already possess, I am quite good at teaching myself the new skills or finding someone to help me learn them.

strongly agree __ agree__ disagree__ strongly disagree __

14. The district offers me a wide and rich menu of learning opportunities that allow me to match my preferred learning styles with the activities I select.

strongly agree __ agree__ disagree__ strongly disagree __

15. Not enough time is made available for me to figure out smart ways to use technology and build the lessons and units that would comfortably blend such tools into the work of my classes.

strongly agree __ agree__ disagree__ strongly disagree ____

16. I am making more time now than I used to for students to do more of the thinking - analyzing, interpreting, inferring and synthesizing.

strongly agree __ agree__ disagree__ strongly disagree __

17. I would do more with new technologies if it were not for the pressures that are loaded onto me by the new state standards and tests.

strongly agree __ agree__ disagree__ strongly disagree __

18. If I get stuck or frustrated with something new, whether it be technology or some other aspect of teaching and learning, I know whom to turn to if I want support and assistance.

strongly agree __ agree__ disagree__ strongly disagree __

Index